Groundwater Transport:

Handbook of Mathematical Models

WATER RESOURCES MONOGRAPH SERIES **10**

Groundwater Transport:

Handbook of Mathematical Models

Iraj Javandel, Christine Doughty,
and Chin-Fu Tsang

AMERICAN GEOPHYSICAL UNION
WASHINGTON, D.C.
1984

Groundwater Transport:
Handbook of Mathematical Models

Library of Congress Cataloging in Publication Data

Main entry under title:

Groundwater transport.

(Water resources monograph series ; 10)
Bibliography: p.
1. Water, Underground—Pollution—Mathematical models—Handbooks, manuals, etc. 2. Groundwater flow—Mathematical models—Handbooks, manuals, etc. I. Javandel, Iraj. II. Doughty, Christine. III. Tsang, Chin-Fu. IV. Series: Water resources monograph ; 10.
TD426.G77 1984 628.1'68 84-6452
ISBN 0-87590-313-4

ISSN 0270-9600

Printed in the United States of America.

CONTENTS

PREFACE

Concerns over groundwater pollution have resulted in the passage of legislation during the last decade calling for pollution control and remedial measures to ensure proper drinking water quality. There are two main types of groundwater pollution caused by man: (1) pollution caused by the use of pesticides, herbicides, and fertilizers over agricultural lands, where the source of contamination covers a relatively large area, and (2) pollution caused by industries and municipalities, which is generally more localized. For the second type, because the contamination in the groundwater is localized, the design of any remedial measure requires knowledge of the extent of the contaminant plume. Various mathematical methods may be used for estimating the size, shape, and development of a localized contaminant plume. It is the need for these mathematical methodologies that forms the main impetus for the preparation of this monograph.

The study of solute transport in groundwater is a relatively old subject. Initially, various analytical methods were derived for very simple cases. Then a number of semianalytical methodologies were developed that could be applied with the help of simple computers. More recently, a number of numerical approaches have been used to code sophisticated numerical models that can be used for more complicated situations. The present monograph attempts to put together selected analytical solutions, semianalytical methods, and numerical approaches and discuss their strengths and possible pitfalls in application. Comprehensive tables and computer program listings are included in the appendices. On the one hand, we hope that the monograph can be readily used by groundwater hydrologists to study the extent and development of solute plumes in aquifers. On the other hand, we hope that the monograph also gives a brief overview of the subject to encourage readers to embark on further research to enlarge the mathematical methodologies available for handling this important subject.

The material in this monograph was originally prepared for the U.S. Environmental Protection Agency (EPA), Robert S. Kerr Environmental Research Laboratory (RSKERL), in part pursuant to Interagency Agreement AD 89F 2A 175 between the U. S. EPA and the U. S. Department of Energy and in part under U. S. Department of Energy contract DE-AC03-76SF00098. Jack W. Keeley of RSKERL provided technical guidance during the course of the study and Joseph F. Keely of RSKERL reviewed the manuscript and provided constructive comments. We acknowledge their assistance as well as their encouragement to publish this monograph. Although publication of this monograph has been approved by EPA, their approval does not signify that the contents necessarily reflect the views and policies of EPA, nor does the mention of trade names or commercial products constitute endorsement or recommendation for use.

Finally, the authors gratefully acknowledge J. S. Y. Wang for his assistance in preparing Chapter 4, J. Noorishad for reviewing the manuscript, P. Fuller for calculations and plotting, and L. Armetta, J. Grant, and S. Kerst for manuscript organization and word processing.

I. JAVANDEL
C. DOUGHTY
C. F. TSANG

Earth Sciences Division
Lawrence Berkeley Laboratory
University of California
Berkeley, California 94720

1 Introduction

1.1. Statement of the Problem

Enormous amounts of waste materials, potentially hazardous to groundwater, are stored or disposed of on or beneath the land surface. In many instances, contaminants such as organic and inorganic chemicals and bacteriological substances are found in groundwater, indicating that many of the waste disposal sites communicate with underground water resources.

The overall goal of the Hazardous Substances and Waste Research Program (HSWRP) is to provide the scientific and technical expertise necessary to enable the responsible personnel to discover, control, and clean up hazardous substances and oil that have been released to the environment from various sources. In respect to this overall plan the present handbook attempts to provide a useful guide by which field personnel can become familiar with the state-of-the-art methodology in modeling contaminant transport in the subsurface. This guide will enable users to make initial estimations of contaminant transport at a given site and, if the need arises and the data justify it, to select and to make use of sophisticated numerical models.

1.2. Objective and Approach

The objective of this work is to review, select, compile, and demonstrate some of the best and most usable mathematical methods for predicting the extent of subsurface contamination in a format useful to field response personnel. The methods presented range from simple analytical and semianalytical solutions to complex numerical codes. Detailed discussions of the assumptions underlying application of the methods are given. Primary emphasis is on the use of simple formulas and comprehensive tables so that the handbook is practically oriented and readily usable as a guide in the field.

Three different levels of complexity and sophistication are used to address the prediction of contaminant transport in groundwater. These levels are as follows.

1. Simple analytical methods based on the solution of applicable differential equations are used, making a simplified idealization of the field and giving qualitative estimates of the extent of contaminant transport.

2. Semianalytical methods based on the concept of complex velocity potential are used, providing the streamlines for steady state fluid flow and the corresponding contaminant movement in the presence of an arbitrary number of sources and sinks. An average geological environment is assumed and a schematic chemical retardation factor is considered.

3. Sophisticated numerical models are used, accounting for complex geometry and heterogenous media, as well as dispersion, diffusion, and chemical retardation processes (e.g., sorption, precipitation, radioactive decay, ion exchange, degradation).

At the first two levels, appropriate methods are given, computer program listings and their user's guides are attached, and comprehensive tables and figures are presented. For the third level, different numerical approaches are introduced, and a number of presently available numerical codes are tabulated, based on recent surveys by various groups. These tables include model names, key characteristics, and the personnel involved in their development. As an illustration, one of these sophisticated models is described and an example of its application is demonstrated.

1.3. Sources of Contamination

A report to Congress by the Environmental Protection Agency in 1977 conveyed that over 17×10^6 waste-disposal facilities in the United States are emplacing at least 6.5×10^9 m^3 of contaminated liquid into the ground each year. Although 16.6×10^6 of these sites are domestic septic tanks, they are only responsible for about 3×10^9 m^3 of effluent. The other 400,000 disposal facilities are responsible for the remaining 3.5×10^9 m^3. These facilities, described in detail elsewhere [*U.S. Environmental Protection Agency,* 1977], involve the following: (1) industrial wastewater impoundments, (2) land disposal sites for solid wastes, (3) waste disposal through wells, (4) septic tanks and cesspools, (5) collection, treatment, and disposal plants for municipal wastewater, (6) land spreading of sludges, (7) brine disposal from petroleum exploration and development, (8) disposal of mine wastes, (9) agricultural land leachate, (10) chemical spills, and (11) leaks from underground chemical storage facilities. In the following section, we shall briefly discuss some of the important sources of the groundwater contamination listed above.

1.3.1. *Industrial Wastewater Impoundments*

Industrial wastewater impoundments are natural or artificial depressions in the ground used for the temporary or permanent storage and/or disposal of liquid wastes. The surface area of these impoundments varies from a few square meters to several hectares. Their depth is generally small to improve evaporation efficiency. It has been reported [*Josephson,* 1982] that about 70% of the industrial impoundments have no impermeable liner. Therefore the hazardous waste can easily infiltrate downward toward underlying groundwaters. For those which are lined, the general rule has been to design the coefficient of permeability of the liner to be less than 10^{-7} cm/s. The corresponding rate of leakage for such liners is about 31,000 m^3/yr from each square kilometer of the impoundment surface. It has been reported [*Anderson,* 1982] that some of the organic chemicals contained in such impoundments destructively increase the permeability of the clay liners, leading to much greater leakage from the impoundments.

Many different potentially hazardous substances are available from industrial waste and are discharged into these impoundments. Table 1 presents a list of components of wastewater from different industries having significant potential for pollution of groundwater [*U.S. Environmental Protection Agency,* 1973]. Included among the potential contaminants are chlorinated solvents, arsenic, mercury, lead, cyanide, chromium, uranium, and many other toxic organic and inorganic materials. Since the maximum allowable concentration of these substances in drinking water is often on the order of a few parts per billion (ppb), it is obvious that even very slight leakage from industrial wastewater impoundments can lead to serious incidents of groundwater contamination.

TABLE 1. Industrial wastewater components having or indicating significant groundwater contamination potential

Pulp and Paper Industry			
Ammonia	Heavy metals	pH	TDS
COD	Nutrients	Phenols	TOC
Color	(nitrogen	Sulfite	
	and phosphorus)		
Petroleum Refining Industry			
Ammonia	Cyanide	Odor	TDS
Chloride	Iron	pH	TOC
Chromium	Lead	Phenols	Total Phosphorus
COD	Mercaptans	Sulfate	Turbidity
Color	Nitrogen	Sulfide	Zinc
Copper			
Steel Industries			
Ammonia	Cyanide	Phenols	Tin
Chloride	Iron	Sulfate	Zinc
Chromium	pH		
Organic Chemicals Industry			
COD	pH	TDS	Total Nitrogen
Cyanide	Phenols	TOC	Total Phosphorus
Heavy metals			
Inorganic Chemicals, Alkalies, and Chlorine Industry			
Acidity/	Chlorinated	Fluoride	Sulfate
Alkalinity	Benzenoids	Iron	TDS
Aluminum	and Polynuclear	Lead	Titanium
Arsenic	Aromatics	Mercury	TOC
Boron	Chromium	Phenols	Total Phosphorus
Chloride	Cyanide		
	COD		
Plastic Materials and Synthetics Industry			
Ammonia	COD	Organic	Phosphorus
Chlorinated	Cyanide	Nitrogen	Sulfate
Benzenoids	Mercaptans	pH	TDS
and Polynuclear	Nitrate	Phenols	Zinc
Aromatics			
Nitrogen Fertilizer Industry			
Ammonia	COD	pH	Sulfate
Calcium	Iron, Total	Phosphate	TDS
Chloride	Nitrate	Sodium	Zinc
Chromium	Organic Nitrogen		
Phosphate Fertilizer Industry			
Acidity	Fluoride	Nitrogen	Sulfate
Aluminum	Iron	pH	TDS
Arsenic	Mercury	Phosphorus	Uranium
Calcium			

COD Carbon oxygen demand
TOC Total organic carbon
TDS Total dissolved solids
From *U. S. Environmental Protection Agency* [1973]

1.3.2. *Land Disposal of Solid Wastes*

Solid waste land disposal occurs as a result of several types of operations: dumps, landfills, sanitary landfills, and secured landfills [*U.S. Environmental Protection Agency*, 1977]. A dump is an uncovered disposal site where solid or liquid wastes are deposited. If the wastes are periodically covered with natural soils, a landfill is created. Sanitary landfills are sites where solid wastes are disposed of by compacting the waste and covering it at the end of each operating day to minimize environmental hazards. If efforts are made to prohibit contaminant movement between the waste and the surrounding environment (particularly the groundwater), it is called a secured landfill [*Farb*, 1978]. Secured landfills are generally designed to accept highly toxic waste and are supposed to be continuously monitored.

Disposing of waste in dumps and landfills is a very common practice while the use of true sanitary landfills is rare [*U.S. Environmental Protection Agency*, 1977]. Secured landfills are still at the experimental stage. Four out of five such secured landfills, constructed during recent years in the state of New Jersey, have experienced operational problems [*Montague*, 1982]. It is estimated that about 20,000 land disposal sites accommodate municipal wastes in the United States. Most of these facilities are open dumps that are poorly sited and operated, yet most have received some industrial wastes [*U.S. Environmental Protection Agency*, 1977]. The number of privately owned industrial land disposal sites is not accurately known, but they are suspected to outnumber municipal landfills.

The mechanism of groundwater contamination by solid waste land disposal facilities is mainly through the generation of leachate, with subsequent downward movement to underlying groundwaters. Leachate generation is due in part to precipitation which percolates through the solid waste, dissolves, and carries out the soluble components of the waste. This liquid, together with any liquid waste placed in the fill and other liquids coming from waste decomposition, constitutes the leachate. The volume of leachate generated at each period of time depends on the availability of moisture within the waste. Since the contribution of moisture from precipitation is generally essential in forming leachate, one might expect that a dump located in a humid area would generate the most leachate. Furthermore, the groundwater level in humid areas is generally much shallower than in arid or semiarid locations, which results in a greater risk of groundwater contamination by an uncovered land disposal site in a humid area. Unless a site of this type is properly designed and located, the site receives not only all of the precipitation falling directly on the site but additional inflow from adjacent surface runoff. Since the rate of evaporation in humid areas is relatively low, the available volume of water for leachate generation may be extremely high.

Secured and sanitary landfills, on the other hand, may be designed so that liquid wastes are not allowed, the inflow of surface runoff is not permitted, the site is properly lined, and generated leachate is collected and removed from the site. These features minimize the risk of groundwater contamination. Unfortunately, the vast majority of the land disposal sites in operation do not have these features. For example, of the 18,500 municipal land disposal sites operating in 1974, only about 20 sites were lined and only 60 sites had leachate treatment facilities [*U.S. Environmental Protection Agency*, 1977]. In regard to industrial land disposal sites, little information is available due to restricted access to sites and records, but they are expected to differ little from municipal sites.

The composition of leachate generated at an individual site is clearly a function of the type of waste deposited in that location. The composition of leachate from some municipal waste disposal sites has been investigated and reported by the *U.S. Environmental Protection Agency* [1974]. The concentration of some of the major components in 20 samples of leachate from municipal solid wastes indicated a variation of 3 to 5 orders of magnitude, emphasizing the fact that each site should be studied individually. *Anderson* [1982] published a typical description of the contents of drums placed in industrial landfills in the State of New York. *Robertson et al.* [1974] identified more than 40 organic compounds in leachate-contaminated groundwater in an aquifer in Oklahoma.

A case where leachate migration caused serious pollution of a large aquifer used as a city's water supply is described by *Apgar and Satherwaite* [1975]. *Kimmel and Braids* [1974] delineated a leachate plume that is more than 3000 m long and greater than 50 m in depth at a landfill overlying sand and gravel on Long Island, New York.

A partial listing, prepared by the U.S. Environmental Protection Agency, of the potentially hazardous constituents available in the waste of some industries is given in Table 2.

1.3.3. *Waste Disposal Through Wells*

In general, two types of wells are used for injection of liquid waste into subsurface strata, shallow wells and deep wells. Storm water, spent cooling water, and sewage effluent are generally injected through relatively shallow wells. Sometimes these wells are completed in the unsaturated zone; however, they often penetrate the saturated zone and thus lead the recharging liquid directly into the groundwater. Tens of thousands of these shallow wells operate throughout the United States.

Large volumes of brine produced by petroleum industries, geothermal energy production, and other sources are generally injected through deep wells (ranging in depth between approximately 300 to 2000 m) into saline water aquifers. Hazardous chemicals, petrochemicals, and pharmaceuticals are also injected.

The use of wells for injection of either sewage or industrial wastes into freshwater aquifers is forbidden in the United States. However, a number of documented cases of severe groundwater contamination resulting from the illegal disposal of hazardous wastes into wells have been reported [*U.S. Environmental Protection Agency,* 1977]. In a few areas, principally in limestone and basalt regions where openings in the rock are large enough to transmit high volumes of liquid, the practice of discharging raw sewage and industrial wastes into shallow freshwater aquifers is not uncommon. Although the volume of industrial wastes injected into the subsurface is very small relative to other types of injected wastes, the extremely hazardous nature of these injectants requires strict regulations and monitoring.

It was formerly believed that, due to adsorption and biodegradation, injected wastewater became free of contamination after passing a short distance through the porous medium. Recently, however, it has been shown that some of the organic chemicals and viruses can migrate long distances through aquifers before being completely eliminated. The contamination of fresh groundwater by deep injection may occur through the following mechanisms: (1) the increase of pressure within deep aquifers due to the addition of injectants may facilitate the upward movement of toxic materials through abandoned deep wells and undetected conductive faults and fracture zones, (2) upward movement of the waste from the saline aquifer along the outside of the well casing, and (3) leakage through the confining beds due to unplanned hydraulic fracturing.

TABLE 2. Components of industrial waste

	Metals mining	Primary metals	Pharmaceuticals	Batteries	Inorganic chemicals	Organic chemicals	Pesticides	Explosives	Paints	Petroleum refining	Electroplating
Ammonium salts		X								X	
Antimony	X				X				X		
Arsenic	X	X	X		X					X	
Asbestos					X				X		
Barium									X		
Beryllium	X									X	
Biological waste			X								
Cadmium	X	X		X	X				X	X	X
Chlorinated hydrocarbons					X	X			X		X
Chromium		X	X	X	X				X	X	X
Cobalt									X	X	
Copper	X	X	X	X					X	X	X
Cyanide		X			X					X	X
Ethanol waste, aqueous			X								
Explosives (TNT)								X			
Flammable solvents						X			X		
Fluoride		X			X						
Halogenated solvents			X								
Lead solvents	X	X		X	X				X	X	X
Magnesium	X										
Manganese		X									
Mercury		X	X	X	X				X	X	
Molybdenum										X	
Nickel		X		X	X					X	
Oil		X								X	X
Organics, misc.						X					
Pesticides (organophosphates)							X				
Phenol		X								X	X
Phosphorus					X						X
Radium	X										
Selenium	X	X	X							X	
Silver				X						X	X
Vanadium										X	
Zinc	X	X	X	X	X				X	X	X

From *U. S. Enviromental Protection Agency* [1977]

1.4. Organization of This Handbook

This handbook consists of three major sections, corresponding to three levels of complexity. The sections cover (1) analytical, (2) semianalytical, and (3) numerical methods of predicting the extent of subsurface contamination, respectively. Each section consists of a brief description of the background and fundamental theory of the method, followed by solutions to several selected mathematical models. The advantages and limitations of each method are discussed at the end of each section. Simple computer programs (including a user's guide for each) are listed in appendices for the evaluation of analytical solutions. Tabulated results for a wide range of parameters are also provided. Field-oriented examples are used to illustrate the proper application of these solutions.

The section on semianalytical methods addresses three types of field problems. The first studies field problems that include arbitrary numbers of recharge and discharge wells tapping an aquifer having a pronounced uniform flow field. The second expands the applicability of the first procedure to problems including circular recharge ponds with finite radii. The use of a production well for mapping a contaminant concentration distribution within an aquifer is demonstrated in the third type of problem. A complete set of computer programs and user's guides is provided to enable easy application of these methods. Application of these models is demonstrated through several problems with various degrees of complexity.

Under numerical methods, various approximation techniques are discussed, a list of available models is presented, and the use of one of these models is demonstrated. In the subsequent discussion section, emphasis is placed on the selection of the proper model for a particular problem.

2 Analytical Methods

Analytical methods that handle solute transport in porous media are relatively easy to use. However, because of the complexity of the equations involved, the analytical solutions available are generally restricted to either radial flow problems or to cases where velocity is uniform over the area of interest. Numerous analytical solutions are available for time-dependent solute transport within media having steady and uniform flow. In the following sections, governing equations for such problems will be reviewed (section 2.1), some one- and two-dimensional solutions will be described (sections 2.2 and 2.3), application of these solutions will be illustrated through simple examples (section 2.4), and finally, the advantages and limitations of analytical methods will be discussed (sections 2.5 and 2.6).

2.1. Governing Equations

The partial differential equation describing solute transport is usually written as

$$\frac{\partial}{\partial x_i}\left(D_{ij}\frac{\partial C}{\partial x_j}\right) - \frac{\partial}{\partial x_i}(Cv_i) - \frac{C'W^*}{n} + \sum_{k=1}^{N} R_k = \frac{\partial C}{\partial t} \tag{1}$$

where

$$v_i = \frac{-K_{ij}}{n}\frac{\partial h}{\partial x_j} \tag{2}$$

and

- C solute concentration;
- v_i seepage or average pore water velocity in the direction x_i;
- D_{ij} dispersion coefficient tensor;
- C' solute concentration in the source or sink fluid;
- W^* volume flow rate per unit volume of the source or sink;
- n effective porosity;
- h hydraulic head;
- K_{ij} hydraulic conductivity tensor;
- R_k rate of solute production in reaction k of N different reactions;
- x_i Cartesian coordinate.

According to *Grove* [1976], if equilibrium-controlled ion exchange reactions are considered, the summation in (1) may be set equal to

$$\sum_{k=1}^{N} R_k = -\frac{\rho_b}{n}\frac{\partial \overline{C}}{\partial t} \tag{3}$$

where ρ_b is the bulk density of the solid and $\overline{C}$ is the concentration of species adsorbed on the solid. To incorporate (3) into (1), an expression relating the adsorbed concentration $\overline{C}$ to the solute concentration C is required. Considering equilibrium transport and assuming that the adsorption isotherm can be described with a linear and reversible equation, one can write

$$\overline{C} = K_d C \tag{4}$$

where K_d is called the distribution coefficient. Now, by incorporating (3) and (4) into (1), we obtain

$$\frac{\partial}{\partial x_i}\left(D_{ij}\frac{\partial C}{\partial x_j}\right) - \frac{\partial}{\partial x_i}(Cv_i) - \frac{C'W^*}{n} = R\frac{\partial C}{\partial t} \tag{5}$$

where

$$R = \left(1 + \frac{\rho_b K_d}{n}\right) \tag{6}$$

The parameter R is called the retardation factor. If the velocity of the contaminant v_c, the groundwater velocity v, ρ_b, and n are known, and if the flow is approximately one-dimensional, K_d can be estimated from the retardation equation [*Davis and DeWiest,* 1966]:

$$R = \frac{v}{v_c} \tag{7}$$

The K_d values have been measured in the laboratory and in the field. For example, *Patterson and Spoel* [1981] used a modified batch method and *Pickens et al.* [1981] used a radial injection dual-tracer test to obtain the distribution coefficient K_d of Sr for some specific aquifers.

If one needs to consider radioactive decay, then the reaction term ΣR_k in (1) should also include an expression dealing with this process. The radioactive decay reaction is expressed as [*Anderson,* 1979]

$$-\lambda\left(C + \frac{\rho_b \overline{C}}{n}\right)$$

where $\lambda = \ell n\ 2/\text{half-life}$ is called the radioactive decay constant. If (4) is used, the above expression for the radioactive decay reaction becomes

$$-\lambda C\left(1 + \frac{\rho_b K_d}{n}\right) = -\lambda CR \tag{8}$$

Adding this expression to (5), one obtains

$$\frac{\partial}{\partial x_i}\left(D_{ij}\frac{\partial C}{\partial x_j}\right) - \frac{\partial}{\partial x_i}(Cv_i) - \frac{C'W*}{n} - \lambda CR = R\frac{\partial C}{\partial t} \quad (9)$$

In general, v_i in (1), (2), (5), and (9) is a function of both time and space. The value of v_i is calculated from (2). Distribution of hydraulic head h at different times should be obtained from the solution of the following equation:

$$\frac{\partial}{\partial x_i}\left(K_{ij}\frac{\partial h}{\partial x_j}\right) = S_s\frac{\partial h}{\partial t} + W* \quad (10)$$

where S_s is specific storage.

An analytical solution to (9) in general form is not tractable. In fact, no analytical or numerical solutions are available for anisotropic systems. This is quite a problem because very often porous media in nature are anisotropic with respect to hydraulic conductivity. Fractured media, unless they are so broken that they can be considered as equivalent porous media, are always anisotropic. Except for some radial flow problems, almost all available analytical solutions belong to systems having a uniform and steady flow. This means that the magnitude and direction of the velocity throughout the system are invariable with respect to time and space, which requires the system to be homogeneous and isotropic with respect to the hydraulic conductivity.

Equations (9), (2) and (10) for homogeneous and isotropic media under steady state uniform flow without considering recharge and discharge become

$$\frac{\partial}{\partial x_i}\left(D_{ij}\frac{\partial C}{\partial x_j}\right) - v_i\frac{\partial C}{\partial x_i} - \lambda CR = R\frac{\partial C}{\partial t} \quad (11)$$

where

$$v_i = -\frac{K}{n}\frac{\partial h}{\partial x_i} \quad (12)$$

$$\frac{\partial^2 h}{\partial x_i{}^2} = 0 \quad (13)$$

Usually, the dispersion coefficient tensor in (11) is further simplified. Dispersion is the result of two processes, molecular diffusion and mechanical mixing. Diffusion is the process whereby ionic or molecular constituents move under the influence of their kinetic activity in the direction of their concentration gradients. Fick's first law, expressed as

$$F = -D^* \frac{dC}{dx} \tag{14}$$

describes the process of molecular diffusion, where F is the mass flux of solute and D^* diffusion coefficient. The diffusion coefficients for electrolytes in aqueous solutions are well known. Values of D^* for major ions may be obtained from *Robinson and Stokes* [1965]. In porous media the effective diffusion coefficient is generally smaller. If the effective diffusion coefficient is shown by $\bar{D}$, then

$$\bar{D} = wD^* \tag{15}$$

where w is a number less than 1 and should be determined empirically. *Perkins and Johnston* [1963] suggested that the value of w is approximately 0.707. *Bear* [1972] suggests that w is equivalent to the tortuosity of the granular medium with a value close to 0.67.

The mechanical mixing component of the dispersion process is the result of velocity variations within the porous medium. *Scheidegger* [1961] assumed this component of the dispersion coefficient to be directly proportional to the seepage velocity and concluded that

$$D_{ij} = \alpha_{ijlm} \frac{v_l v_m}{v} \tag{16}$$

where α_{ijlm} is a fourth rank tensor, v_l and v_m are velocity components, and v is the magnitude of the velocity vector. Equation (16) shows the source of the complexity in the solution of solute transport problems in anisotropic systems. In fact, *De Josselin de Jong* [1972] has shown that for the general anisotropic case the dispersion coefficient is a tensor of infinite rank. For homogeneous, isotropic porous media, if x_1 is in the direction of the velocity vector v, then $v_1 = v$ and $v_2 = 0$, and according to *Bachmat and Bear* [1964],

$$D_{11} = \alpha_L v \tag{17}$$

$$D_{22} = D_{33} = \alpha_T v \tag{18}$$

$$D_{ij} = 0 \quad i \neq j \tag{19}$$

where α_L and α_T are longitudinal and lateral dispersivities, respectively. The value of α_T is usually an order of magnitude smaller than α_L. Finally, hydrodynamic dispersion coefficients may be written as

$$D_L = \bar{D} + \alpha_L v \tag{20}$$

$$D_T = \bar{D} + \alpha_T v \tag{21}$$

where D_L and D_T are dispersion coefficients along and perpendicular to the flow direction, respectively. Equations (20) and (21), which were derived based on the above procedure, are functions of velocity only. Recently, *de Marsily* [1982] claimed that field experiments indicate that D_L and D_T may be time dependent too. There have also been indications that dispersion coefficients are functions of space.

If one ignores the dependency of dispersion coefficients on space and time and further assumes that the x axis is in the direction of the velocity vector, then for two-dimensional problems (11) reduces to

$$D_L \frac{\partial^2 C}{\partial x^2} + D_T \frac{\partial^2 C}{\partial y^2} - v \frac{\partial C}{\partial x} - \lambda RC = R \frac{\partial C}{\partial t} \tag{22}$$

For nonreactive dissolved constituents (22) simplifies to

$$D_L \frac{\partial^2 C}{\partial x^2} + D_T \frac{\partial^2 C}{\partial y^2} - v \frac{\partial C}{\partial x} = \frac{\partial C}{\partial t} \tag{23}$$

The usual treatment for a two-dimensional problem where the flow lines are curved but the magnitude of velocity remains constant along the flow line is to define two curvilinear coordinate directions, S_l and S_t, where S_l is directed along the flow line and S_t is orthogonal to it. Equation (23) then becomes

$$D_L \frac{\partial^2 C}{\partial S_l^2} + D_T \frac{\partial^2 C}{\partial S_t^2} - v_l \frac{\partial C}{\partial S_l} = \frac{\partial C}{\partial t} \tag{24}$$

Finally, the one-dimensional form of the advection-dispersion equation for non-reactive dissolved constituents in saturated, homogeneous, isotropic materials under steady state uniform flow is

$$D \frac{\partial^2 C}{\partial x^2} - v \frac{\partial C}{\partial x} = \frac{\partial C}{\partial t} \tag{25}$$

2.1.1. *Initial and Boundary Conditions*

Appropriate initial and boundary conditions are required to solve any of the time-dependent partial differential equations given above. The initial condition in general form is written as

$$C(x, y, z, t) = f(x, y, z) \qquad t = 0 \tag{26}$$

where $f(x, y, z)$ is a known function. Usually a constant concentration is assumed throughout the domain of interest.

Three types of boundary conditions may be specified depending on physical constraints.

1. Dirichlet boundary conditions prescribe concentration along a portion of the boundary:

$$C = C_o(x, y, z, t) \tag{27}$$

where $C_o(x, y, z, t)$ is a given function of time and space for that particular portion of boundary.

2. Neumann boundary conditions prescribe the normal gradient of concentration over a certain portion of the boundary:

$$\left(D_{ij} \frac{\partial C}{\partial x_j} \right) n_i = q(x, y, z, t) \tag{28}$$

where q is a known function and n_i are directional cosines. For impervious boundaries q becomes zero.

3. Cauchy boundary conditions prescribe concentration and its gradient:

$$\left(D_{ij} \frac{\partial C}{\partial x_j} - v_i C \right) n_i = g(x, y, z, t) \tag{29}$$

where g is a known function. The first term on the left-hand side represents flux by dispersion, and the second term presents the advection effect.

2.2. One-Dimensional Problems

A relatively complete set of one-dimensional analytical solutions for convective-dispersive solute transport equations has been recently published by *Van Genuchten and Alves* [1982]. Here we shall review a few cases having practical applications.

First, let us consider a one-dimensional model consisting of an infinitely long homogeneous isotropic porous medium with a steady state uniform flow with seepage velocity v. We inject a particular chemical from one end of the model for a period of time t_o such that the input concentration varies as an exponential function of time. The value of that chemical concentration at any time t and at a distance x from the injection boundary, allowing for decay and adsorption, may be obtained from the solution of the following set of equations:

$$D \frac{\partial^2 C}{\partial x^2} - v \frac{\partial C}{\partial x} - \lambda R C = R \frac{\partial C}{\partial t} \tag{30}$$

which is a one-dimensional form of (22).

$$C(x, t) = 0 \qquad t = 0 \tag{31}$$

which means that the system is initially free of that chemical.

$$\frac{\partial C\ (x,t)}{\partial x} = 0 \qquad x = \infty \tag{32}$$

which indicates that the concentration gradient at the other end of the model remains unchanged.

$$\left(-D\frac{\partial C}{\partial x} + vC\right)\bigg|_{x=0} = vf(t) \tag{33}$$

where the input concentration $f(t)$ takes the following form:

$$\begin{aligned} f(t) &= C_o \exp(-\alpha t) \qquad 0 < t \leqslant t_o \\ f(t) &= 0 \qquad t > t_o \end{aligned} \tag{34}$$

where C_o and α are constants. Equation (33) shows that the mass flux of the chemical at the injection boundary at any time is equivalent to the total flux of that chemical carried by dispersion and advection.

Using the Laplace transform technique, *Van Genuchten* [1982] solved the above set of equations with the following results:

$$\begin{aligned} C(x,t) &= A(x,t) \qquad 0 < t \leqslant t_o \\ C(x,t) &= A(x,t) - A(x,t-t_o)\exp(-\alpha t_o) \qquad t > t_o \end{aligned} \tag{35}$$

where

$$\begin{aligned} A(x,t) &= C_o \exp(-\alpha t) A_1(x,t) \qquad \alpha \neq \lambda \\ A(x,t) &= C_o \exp(-\alpha t) A_2(x,t) \qquad \alpha = \lambda \end{aligned} \tag{36}$$

$$A_1(x,t) = \frac{v}{v+U}\exp\left[\frac{x(v-U)}{2D}\right]\operatorname{erfc}\left[\frac{Rx - Ut}{2(DRt)^{1/2}}\right]$$

$$+ \frac{v}{v-U}\exp\left[\frac{x(v+U)}{2D}\right]\operatorname{erfc}\left[\frac{Rx + Ut}{2(DRt)^{1/2}}\right]$$

$$+ \frac{v^2}{2DR(\lambda-\alpha)}\exp\left[\frac{vx}{D} + (\alpha-\lambda)t\right]\operatorname{erfc}\left[\frac{Rx + vt}{2(DRt)^{1/2}}\right] \tag{37}$$

with

$$U = [v^2 + 4DR(\lambda - \alpha)]^{1/2} \tag{38}$$

$$A_2(x,t) = \frac{1}{2}\,\mathrm{erfc}\left[\frac{Rx - vt}{2(DRt)^{1/2}}\right] + \left(\frac{v^2 t}{\pi DR}\right)^{1/2} \exp\left[-\frac{(Rx - vt)^2}{4DRt}\right]$$

$$-\frac{1}{2}\left(1 + \frac{vx}{D} + \frac{v^2 t}{DR}\right) \exp\left(\frac{vx}{D}\right) \mathrm{erfc}\left[\frac{Rx + vt}{2(DRt)^{1/2}}\right] \tag{39}$$

A computer program has been provided in Appendix B which enables one to calculate the ratio of C/C_o from (35) through (39) for any given point downstream from the source of contamination and at any given time. Appendix A gives a series of tables listing dimensionless concentration C/C_o as a function of average pore water velocity, dispersion coefficient, retardation factor, decay constant, and the period of activity of the source. These tables have been prepared with the computer program given in Appendix B. An example of the use of the tables and computer program is given in section 2.4.

2.2.1. *Specific Cases*

Solutions for three specific cases are examined below.

1. *For $\alpha = 0$, constant input concentration.* When $\alpha = 0$, the input boundary condition (33) changes to a constant flux as follows:

$$\left(-D\frac{\partial C}{\partial x} + vC\right)\bigg|_{x=0} = vC_o \qquad 0 < t \leqslant t_o$$

$$\left(-D\frac{\partial C}{\partial x} + vC\right)\bigg|_{x=0} = 0 \qquad t > t_o \tag{40}$$

The solution to this problem may be easily obtained by letting α go to zero in (35) through (37). In this case, for $t \leqslant t_o$, one obtains

$$\frac{C}{C_o} = A_1(x,t) = \frac{v}{v + U} \exp\left[\frac{x(v - U)}{2D}\right] \mathrm{erfc}\left[\frac{Rx - Ut}{2(DRt)^{1/2}}\right]$$

$$+ \frac{v}{v - U} \exp\left[\frac{x(v + U)}{2D}\right] \mathrm{erfc}\left[\frac{Rx + Ut}{2(DRt)^{1/2}}\right]$$

$$+ \frac{v^2}{2DR\lambda} \exp\left[\frac{vx}{D} - \lambda t\right] \operatorname{erfc}\left[\frac{Rx + vt}{2(DRt)^{1/2}}\right] \tag{41}$$

with

$$U = (v^2 + 4DR\lambda)^{1/2} . \tag{42}$$

2. *For $\alpha = \lambda = 0$, no decay factor and constant input concentration.* When λ is zero, the governing differential equation (30) becomes

$$D\frac{\partial^2 C}{\partial x^2} - v\frac{\partial C}{\partial x} = R\frac{\partial C}{\partial t} \tag{43}$$

which is valid if the solute is not subject to decay but adsorption still is permitted. If α is also zero then the inflow boundary condition becomes independent of time, as was given by (40). The solution to (43) subject to the initial condition (31) and boundary conditions (32) and (40) may be given by

$$\begin{aligned} \frac{C}{C_o} &= A_2(x, t) \qquad 0 < t \leqslant t_o \\ \frac{C}{C_o} &= A_2(x, t) - A_2(x, t - t_o) \qquad t > t_o \end{aligned} \tag{44}$$

where $A_2(x, t)$ was given by (39).

3. *For $\alpha = \lambda = 0$; $R = 1$.* If, in addition to the conditions of no decay and constant influx of solute, one deals with nonreactive dissolved constituents, i.e., $K_d = 0$ and as such, $R = 1$, the solution can be easily obtained by letting $R = 1$ in (39) and (44):

$$\begin{aligned} \frac{C}{C_o} &= A_3(x, t) \qquad 0 < t \leqslant t_o \\ \frac{C}{C_o} &= A_3(x, t) - A_3(x, t - t_o) \qquad t > t_o \end{aligned} \tag{45}$$

where

$$A_3(x, t) = \frac{1}{2} \operatorname{erfc}\left[\frac{x - vt}{2(Dt)^{1/2}}\right] + \left(\frac{v^2 t}{\pi D}\right)^{1/2} \exp\left[-\frac{(x - vt)^2}{4Dt}\right]$$

$$-\frac{1}{2}\left(1 + \frac{vx}{D} + \frac{v^2 t}{D}\right) \exp\left[\frac{vx}{D}\right] \operatorname{erfc}\left[\frac{x + vt}{2(Dt)^{1/2}}\right] \tag{46}$$

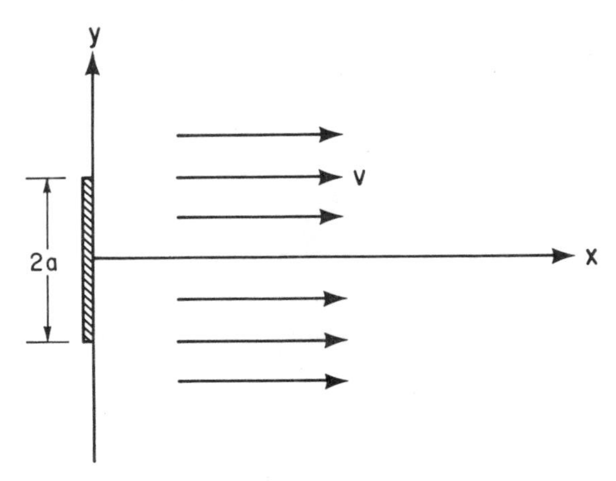

Fig. 1. A schematic diagram showing the two-dimensional plane dispersion model.

As seen in (43), when $\lambda = 0$ the effect of adsorption, as introduced by the retardation factor, can be handled simply by a change of time scale, $T = t/R$. Note that the computer program given in Appendix B encompasses all the above specific cases.

2.3. Two-Dimensional Problems

Several analytical solutions for two-dimensional dispersion problems are available. *Ogata* [1970] and *Cleary and Ungs* [1978] have described some of these solutions, however, many of them do not represent a realistic field problem. Here we shall select and review those solutions which may be useful for estimating the extent of contaminant transport in a groundwater flow system.

2.3.1. *Two-Dimensional Plane Dispersion Model*

Let us consider a homogeneous, isotropic porous medium having a unidirectional steady state flow with seepage velocity v. If we choose a Cartesian coordinate system with x axis oriented along the direction of flow (Figure 1) and if the magnitude of the dispersion coefficients in that direction and orthogonal to it are defined by D_L and D_T, respectively, then the two-dimensional advection-dispersion equation, as given by (22), can be used:

$$D_L \frac{\partial^2 C}{\partial x^2} + D_T \frac{\partial^2 C}{\partial y^2} - v\frac{\partial C}{\partial x} - \lambda RC = R\frac{\partial C}{\partial t}$$

where R is the retardation factor for the given type solute.

Let us assume that initially the medium is free of a particular solute species and at a certain time a strip type source with length $2a$, orthogonal to the groundwater flow direction, is introduced along the y axis. If the concentration of the solute diminishes exponentially with time, the initial and boundary conditions of this mathematical model may be written as

$$\begin{aligned} C(0, y, t) &= C_o e^{-\alpha t} \qquad -a \leqslant y \leqslant a \\ C(0, y, t) &= 0 \qquad \textit{other values of } y \end{aligned} \tag{47}$$

$$\lim_{y \to \pm\infty} \frac{\partial C}{\partial y} = 0$$

$$\lim_{x \to \infty} \frac{\partial C}{\partial x} = 0$$

An analytical solution [*Cleary and Ungs,* 1978] to the above model may be presented as

$$C(x, y, t) = \frac{C_o x}{4(\pi D_L)^{1/2}} \exp\left(\frac{vx}{2D_L} - \alpha t\right)$$

$$\cdot \int_0^{t/R} \exp\left[-\left(\lambda R - \alpha R + \frac{v^2}{4D_L}\right)\tau - \frac{x^2}{4D_L \tau}\right]\tau^{-3/2}$$

$$\cdot \left[\operatorname{erf}\left(\frac{a - y}{2(D_T \tau)^{1/2}}\right) + \operatorname{erf}\left(\frac{a + y}{2(D_T \tau)^{1/2}}\right)\right] d\tau \qquad (48)$$

A computer program has been provided in Appendix D which enables one to calculate the ratio of C/C_o from (48) for any given point downstream from the source of contamination at any given time. Appendix C gives a series of tables prepared with the computer program, listing dimensionless concentration C/C_o as a function of time, average pore water velocity, half length of the source, longitudinal and transverse dispersion coefficients, and other pertinent parameters. An example of the use of the tables is given in section 2.4.

The solution given above corresponds to the case where the source is orthogonal to the flow direction. If, however, the direction of flow forms an angle with the x axis (Figure 2), then the solution for this case [*Cleary and Ungs,* 1978] may be written as

$$C(x, y, t) = \frac{C_o x}{4(\pi D_x)^{1/2}} \exp\left(\frac{v_x x}{2D_x} - \alpha t\right)$$

$$\cdot \int_0^{t/R} \exp\left[-\left(\lambda R - \alpha R + \frac{{v_x}^2}{4D_x}\right)\tau - \frac{x^2}{4D_x \tau}\right]\tau^{-3/2}$$

$$\cdot \left\{\operatorname{erf}\left[\frac{a - y}{2(D_y \tau)^{1/2}} + \frac{v_y}{2}\left(\frac{\tau}{D_y}\right)^{1/2}\right]\right.$$

$$\left. + \operatorname{erf}\left[\frac{a + y}{2(D_y \tau)^{1/2}} - \frac{v_y}{2}\left(\frac{\tau}{D_y}\right)^{1/2}\right]\right\} d\tau \qquad (49)$$

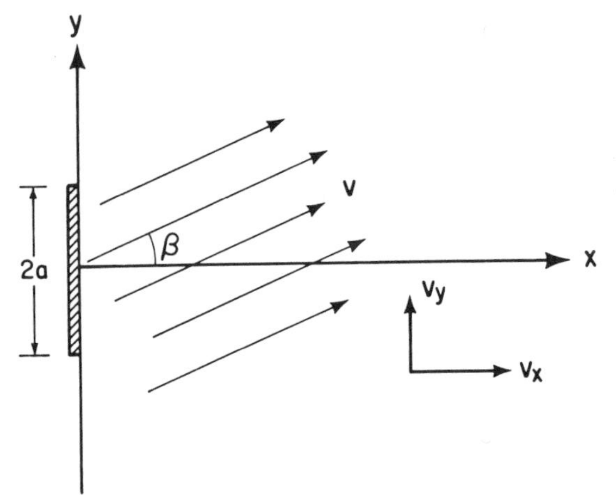

Fig. 2. A schematic diagram showing the two-dimensional plane dispersion model with uniform flow making an angle β with the positive x axis.

where

v_x, v_y components of average pore water velocity along x and y axis, respectively;

D_x coefficient of dispersion along x axis;

D_y coefficient of dispersion along y axis.

2.3.2. *Dispersion in Radial Flow*

Sometimes it is of interest to estimate the extent of contamination within an aquifer when a solute with concentration C_o has been continuously recharged into the aquifer through a fully penetrating well. The following solution can be used to estimate the concentration distribution, provided that the rate of recharge and concentration of recharge fluid remain constant.

The advection-dispersion equation for plane radial flow [*Bear,* 1979] may be written as

$$\frac{1}{r}\frac{\partial}{\partial r}\left(Dr\frac{\partial C}{\partial r}\right) - v\frac{\partial C}{\partial r} = \frac{\partial C}{\partial t} \tag{50}$$

Recalling the definition of dispersion coefficient D and assuming that the molecular diffusion is negligible, one can substitute $\alpha_L v$ for D in (50) to obtain

$$\frac{1}{r}\frac{\partial}{\partial r}\left(\alpha_L vr\frac{\partial C}{\partial r}\right) - v\frac{\partial C}{\partial r} = \frac{\partial C}{\partial t} \tag{51}$$

For cases such as steady plane radial flow where *(vr)* remains constant, (51) can be simplified to the following form:

$$\alpha_L v\frac{\partial^2 C}{\partial r^2} - v\frac{\partial C}{\partial r} = \frac{\partial C}{\partial t} \tag{52}$$

Let us now consider a confined aquifer with thickness b being recharged through a fully penetrating well at a constant rate Q. If the concentration of a particular chemical in the recharge fluid is C_o and the concentration of that chemical in the aquifer water was originally zero, then mathematically this problem may be expressed as follows.

Let us introduce the following dimensionless parameters:

$$r_D = r/\alpha_L \qquad t_D = \frac{Qt}{2\pi b n \alpha_L^2} \qquad C_D = C/C_o$$

where r_D, t_D, and C_D are dimensionless radius, dimensionless time, and dimensionless concentration, respectively. Equation (52) in terms of these dimensionless parameters becomes

$$\frac{1}{r_D}\frac{\partial^2 C_D}{\partial r_D^2} - \frac{1}{r_D}\frac{\partial C_D}{\partial r_D} = \frac{\partial C_D}{\partial t_D} \tag{53}$$

Initial and boundary conditions for this problem are

$$C_D(r_D, t_D) = 0 \qquad t_D = 0$$

$$C_D(r_{Dw}, t_D) = 1 \tag{54}$$

$$\lim_{r_D \to \infty} C_D(r_D, t_D) = 0$$

where r_{Dw} is dimensionless well diameter.

A solution to the above problem has been given by *Ogata* [1958, 1970]. The numerical evaluation of that solution is cumbersome and has not been done.

Recently, *Moench and Ogata* [1981] rewrote the solution of the above problem in the Laplace transform domain, in terms of Airy functions, as

$$\mathscr{C}_D = \frac{1}{s}\exp\left[\frac{r_D - r_{Dw}}{2}\right]\frac{\mathrm{Ai}(Y)}{\mathrm{Ai}(Y_o)} \tag{55}$$

where

$\mathscr{C}_D$ Laplace transform of dimensionless concentration;
s Laplace transform parameter;
$\mathrm{Ai}(Y)$ Airy function;
$Y = s^{-2/3}(sr_D + 1/4)$;
$Y_o = s^{-2/3}(sr_{Dw} + 1/4)$.

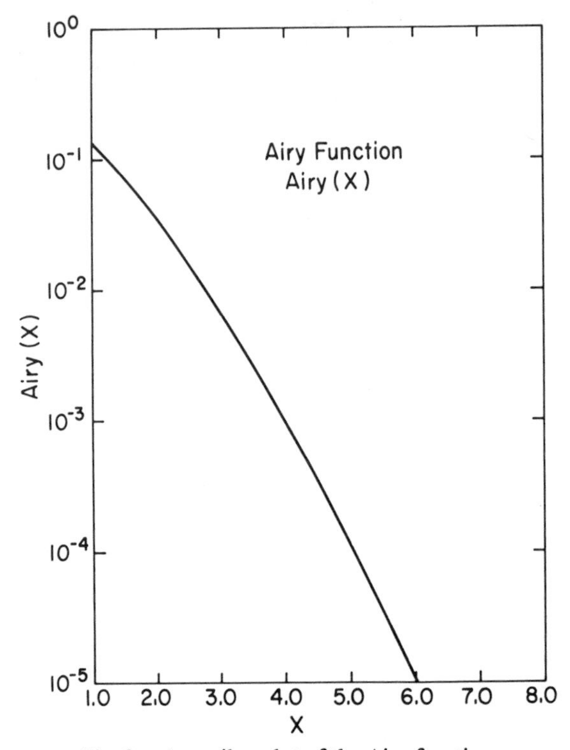

Fig. 3. A semilog plot of the Airy function.

The general form of the Airy function can be found in the work of *Abramowitz and Stegun* [1964]. An asymptotic expansion of this function for large modulus $|Z|$ may be written as

$$\mathrm{Ai}(Z) \simeq \frac{1}{2}\pi^{-1/2}Z^{-1/4}e^{-\xi}\sum_{k=0}^{\infty}(-1)^k c_k \xi^{-k} \qquad |\arg Z| < \pi \tag{56}$$

where

$$c_k = \frac{(2k+1)(2k+3)\cdots(6k-1)}{216^k k!}$$

$$c_o = 1 \qquad \xi = \frac{2}{3}Z^{3/2}$$

Figure 3 shows that the shape of this function for positive arguments is a smooth, monotonic, positive function which tends rapidly to zero. These condi-

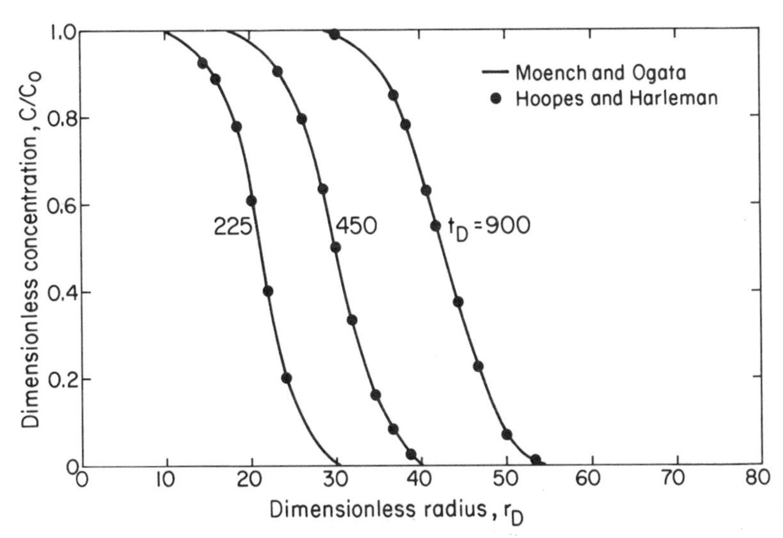

Fig. 4. A comparison of the results of Moench and Ogata's approximate method with *Hoopes and Harleman's* [1967] finite difference scheme, [modified from *Moench and Ogata,* 1981].

tions are generally ideal for the numerical inversion of the Laplace transform. *Moench and Ogata* [1981] applied a method of numerical inversion of the Laplace transform used by *Stehfest* [1970] to evaluate (55). A comparison of their results to those obtained by *Hoopes and Harleman* [1967] using a finite difference technique is given in Figure 4. The computer code used by *Moench and Ogata* [1981] to evaluate (55) is available from the authors upon request. A modified version appears in Appendix F. Appendix E presents values of dimensionless concentration C/C_o as a function of dimensionless time t_D and dimensionless radius r_D for r_{Dw} equal to 1. This table has been prepared with the code given in Appendix F. The value of C/C_o is insensitive to changes in r_{Dw} between 0.1 and 1.0.

2.3.3. *Approximate Solution to Dispersion in Radial Flow*

As was noted, evaluation of the solution proposed by *Moench and Ogata* [1981] requires a computer code. Here, we present another approximate solution [*Raimondi et al.,* 1959; *Hoopes and Harleman,* 1967] which can be easily evaluated with the help of the table of the error function given in Appendix G. This method assumes that at some distance from the source the influence of dispersion and diffusion on the concentration distribution is minimal in comparison to the total dispersion and diffusion that has taken place up to that point.

If one adds the effect of molecular diffusion to the advection-dispersion equation (52) for steady plane radial flow, one obtains

$$\alpha_L v \frac{\partial^2 C}{\partial r^2} - v \frac{\partial C}{\partial r} + \frac{\overline{D}}{r} \frac{\partial}{\partial r} \left(r \frac{\partial C}{\partial r} \right) = \frac{\partial C}{\partial t} \tag{57}$$

where $\overline{D}$ is the coefficient of molecular diffusion. Using the assumption expressed above, (57) leads to the expression

$$\left(\frac{\alpha_L}{v} + \frac{\overline{D}}{v^2} \right) \frac{\partial^2 C}{\partial t^2} - v \frac{\partial C}{\partial r} = \frac{\partial C}{\partial t} \tag{58}$$

For continuous injection of a substance at a steady rate Q with a concentration C_o at $r = 0$, the solution to (58) is given by

$$\frac{C}{C_o} = \frac{1}{2} \operatorname{erfc} \left[\left(\frac{r^2}{2} - rvt \right) \left(\frac{4}{3} \alpha_L r^3 + \frac{\overline{D}}{v} r^3 \right)^{-1/2} \right] \tag{59}$$

where erfc $(x) = 1 - \operatorname{erf}(x)$, and erf (x) and erfc (x) are error function and complementary error functions, respectively. Appendix G gives a table of the error function. Equation (59) was derived based on the initial condition

$$\frac{\partial C(r, 0)}{\partial t} = 0$$

This assumption is approximately true away from the immediate vicinity of the source; however, it is not true very close to the source. Assuming that $\partial C / \partial t = 0$ at $t = 0$, the approximate solution (59) predicts a finite amount of mass in the medium at $t = 0$. Setting $\overline{D} = 0$, *Hoopes and Harleman* [1967] wrote (59) in terms of dimensionless parameters as

$$\frac{C}{C_o} = \frac{1}{2} \operatorname{erfc} \left[\left(\frac{r_D^2}{2} - t_D \right) \left(\frac{4}{3} r_D^3 \right)^{-1/2} \right] \tag{60}$$

where

$$r_D = r / \alpha_L \qquad t_D = \frac{Qt}{2\pi b n \alpha_L^2}$$

and b is the aquifer thickness.

Hoopes and Harleman [1967] have compared the result of (60) with their finite difference solution. They concluded that the results from (60), at any t_D, predict that the substance has moved farther into the medium than is predicted by the finite difference solution. In other words, (60) gives conservative answers. As t_D increases, the approximate solutions (59) and (60) give a better approximation of the concentration distribution. For $t_D > 1000$, the time error between the approximate and finite difference solutions is less than 1%. An example of the use of (60) is given in section 2.4.

2.4. Applications

In this section we present several examples for the analytical solutions given above. The examples have been designed to illustrate the way each type of solu-

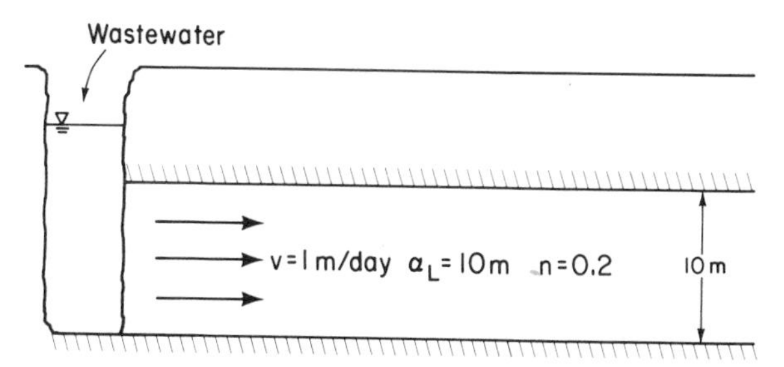

Fig. 5. A schematic diagram showing a vertical section of the aquifer along the direction of flow.

tion may be applied in the field. Because of the inherent limitations of the analytical solutions, some aspects of these examples, particularly the way in which the contaminant is introduced to the aquifer, may not seem to realistically duplicate actual field situations. However, in the absence of sufficient data and other uncertainties, these solutions can be valuable tools for estimating the extent of contamination in the subsurface.

2.4.1. *Example 1*

This example shows how the one-dimensional analytical solutions given in section 2.2, their related tables of dimensionless concentration C/C_o (Appendix A), and the computer code ODAST (Appendix B) can be used. Let us consider a shallow, homogeneous, isotropic aquifer with a thickness of 10 m and a steady uniform average pore water velocity of 1 m/d. A relatively long ditch cuts through this aquifer perpendicular to the direction of flow (Figure 5). A nonreactive chemical waste is being continuously poured into the ditch. The rate of chemical waste inflow is about 0.1 m^3/d per unit length of the ditch. The concentration of a certain nonreactive constituent in this waste is 10 kg/m^3. A longitudinal dispersivity of 10 m and porosity of 0.2 are assumed. Given these conditions, we will perform the following steps: (1) plot the variation of C/C_o versus distance for 1, 2, and 10 years after the start of the operation, and (2) determine how far downstream from the ditch a concentration of 0.1 ppm or more can be found after 10 years.

Multiplying together pore water velocity 1 m/d, aquifer thickness 10 m, porosity 0.2, and unit ditch length 1 m, gives the volumetric rate of groundwater flow from a unit length of the ditch: 2 m^3/d.

The rate of recharge of chemical waste, 0.1 m^3/d, is 5% of the rate of groundwater flow, so one may ignore the increase in groundwater velocity due to chemical waste without introducing a significant error.

Assuming that the mixing of the waste and the natural groundwater in the ditch is perfect, the concentration of the particular solute of interest in the groundwater at the ditch would be calculated as follows. The mass inflow rate of that constituent is 1 kg/d per unit length of the ditch. The volume rate of water leaving each unit length of the ditch is 2 m^3/d. Therefore concentration of the solute in question is $C_o = 1/2$ kg/m^3 or 500 ppm.

The dispersion coefficient D may be estimated from (17) to be 10 m^2/d. The dimensionless concentration C/C_o for $R = 1$, $\alpha = \lambda = 0$, $v = 1$ m/d and $D = 10$ m^2/d for distances up to 1000 m and times of 1, 2, and 10 years is given in the tables in Appendix A. Figure 6 shows plots of dimensionless concentration C/C_o versus distance x for the elapsed times of 1 and 2 years after the start of operation. The above table shows that after 10 years of operation C/C_o is equal to unity up

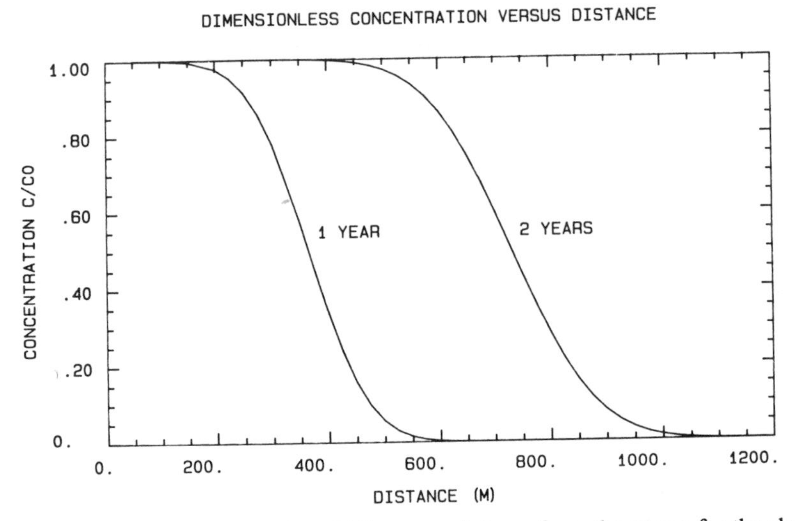

Fig. 6. Dimensionless concentration C/C_o versus distance from the source for the elasped time of 1 or 2 years.

to a distance of x = 1000 m. Thus, in order to plot C/C_o versus x for t = 10 years, we need to investigate distances beyond 1000 m, which will require the computer program given in Appendix B. The input data (Table 3) needed for this program consists of five cards which have been prepared based on the user's guide given in Appendix B.

TABLE 3. Input data for program ODAST for example 1

Card	Column	Entry	Description
1	4-5	12	Total number of x positions
	10	1	Total number of time points
2	1-10	1000.	Distance to the 1st point
	11-20	1500.	Distance to the 2nd point
	21-30	2000.	Distance to the 3rd point
	31-40	2500.	Distance to the 4th point
	41-50	3000.	Distance to the 5th point
	51-60	3500.	Distance to the 6th point
	61-70	4000.	Distance to the 7th point
	71-80	4500.	Distance to the 8th point
3	1-10	5000.	Distance to the 9th point
	11-20	5500.	Distance to the 10th point
	21-30	6000.	Distance to the 11th point
	31-40	6500.	Distance to the 12th point
4	1-10	10.	Time in years
5	1-10	10.	Dispersion coefficient in m^2/d
	11-20	1.	Seepage velocity in m/d
	21-30	1.	Retardation factor
	31-40	20.	Period of solute recharge in years, any value greater than 10 years
	41-50	0.	Solute decay constant
	51-60	0.	Source decay constant

TABLE 4. Dimensionless concentrations C/C_o for distances ranging between 1000 and 6500 m and t = 10 years, calculated for example 1

x, m	C/C_o	x, m	C/C_o
1000	.1000D+01	4000	.9898D−01
1500	.1000D+01	4500	.8471D−03
2000	.1000D+01	5000	.3010D−06
2500	.1000D+01	5500	.3913D−11
3000	.9922D+00	6000	.0000D−13
3500	.7139D+00	6500	.0000D−13

Note: The format .1000D+01 used in this and some other tables stands for $.1000 \times 10^{+01}$.
Values of parameters: v = 1 m/d, D = 10 m^2/d, R = 1, $\lambda = \alpha = 0$, t_o = 20 years.

The output of this program for the above data is shown in Table 4. Note that the major variation of C/C_o occurs between x = 3000 and 4500 m. Therefore, in order to get a more accurate distribution in this range, we run the program for shorter intervals of 100 m; the results are shown in Table 5. Figure 7 shows a plot of variation of C/C_o versus distance from the source for 10 years after the start of operation. In order to determine the second part of the example, note that concentration of 0.1 ppm in this example corresponds to

$$\frac{C}{C_o} = \frac{0.1}{500} = 0.2 \times 10^{-3}$$

Table 4 shows that the values of C/C_o at distances of 4500 and 5000 m are 0.84×10^{-3} and 0.30×10^{-6}, respectively. Thus the concentration ratio of $C/C_o = 0.2 \times 10^{-3}$ should occur somewhere between 4500 and 5000 m from the source.

One may note that in the above calculation the effect of molecular diffusion is ignored. This can be done because the values of $\bar{D}$ in (20) and (21) are generally on the order of 10^{-5} m^2/d which can be safely neglected in comparison with the magnitude of $v\alpha_L$ (10 m^2/d).

2.4.2. *Example 2*

A more realistic case is where the length of the source is finite. When the source can be approximated as a finite length strip, the two-dimensional plane solution

TABLE 5. Dimensionless concentrations C/C_o for shorter intervals between 3100 and 4300 m and t = 10 years, calculated for example 1

x, m	C/C_o	x, m	C/C_o
3100	.9797D+00	3700	.4301D+00
3200	.9532D+00	3800	.2923D+00
3300	.9042D+00	3900	.1796D+00
3400	.8252D+00	4100	.4868D−01
3500	.7139D+00	4200	.2127D−01
3600	.5771D+00	4300	.8230D−02

Values of parameters: v = 1 m/d, D = 10 m^2/d, R = 1, $\lambda = \alpha = 0$, t_o = 20 years.

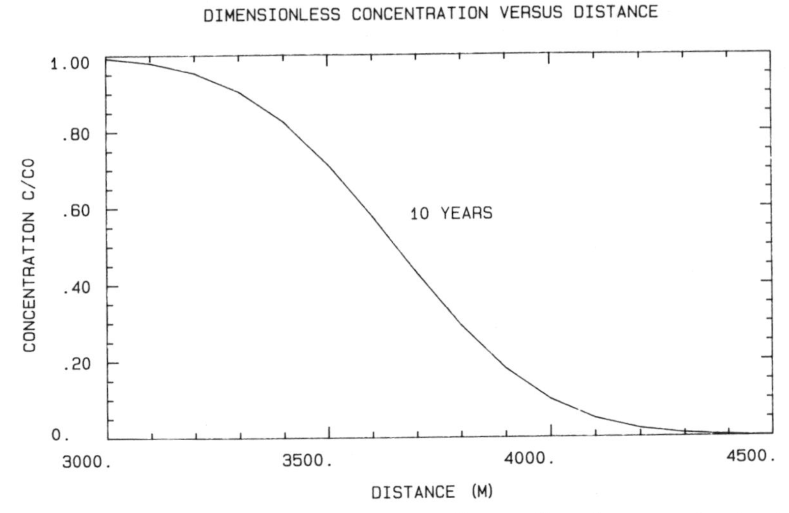

Fig. 7. Dimensionless concentration C/C_o versus distance from the source for the elasped time of 10 years.

presented in section 2.3.1 is a more suitable approach. This example is intended to show how the solution given in section 2.3.1, together with its related tables of values, can be applied in the field.

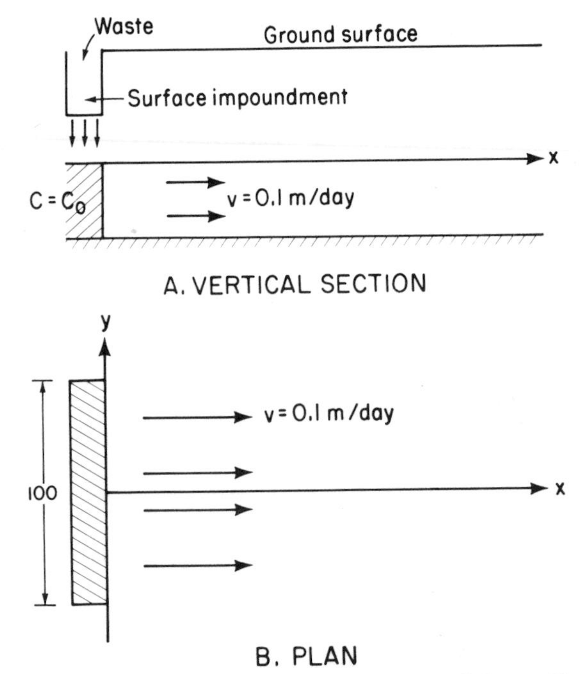

Fig. 8. A schematic diagram showing (*a*) a vertical section of the aquifer and the surface impoundment, (*b*) a plan view of the flow and the source of contamination.

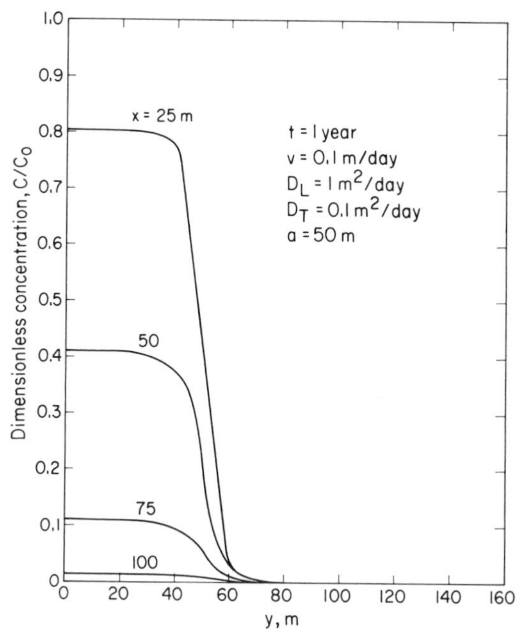

Fig. 9. Dimensionless concentration C/C_o versus distance from the x axis for values of x = 25, 50, 75, and 100 m downstream from the source and t = 1 year, v = 0.1m/d, D_L = 1 m²/d, D_T = 0.1 m²/d, a = 50 m.

Let us consider a relatively thin, shallow, homogeneous, isotropic aquifer having a steady, uniform seepage velocity of 0.1 m/d. Liquid waste from a factory is being discharged into a surface impoundment 100 m long and 5 m wide (Figure 8). For the sake of simplicity let us assume that the entire length of the impoundment ditch is perpendicular to the direction of groundwater flow. Let us suppose that the waste liquid seeping from the bottom of this impoundment reaches the aquifer and creates a constant concentration of 1000 ppm of a certain solute species in the area beneath the ditch. Assuming that the transverse dispersivity of the aquifer is about 1/10 the longitudinal value, we are interested in (1) estimating the variation of concentration downstream from the source 1 and 5 years after the contaminant reaches the aquifer, and (2) given an allowable solute concentration for drinking water of 10 ppm, indicating the area of the aquifer downstream from the source where groundwater is considered to be contaminated 5 years after the solute reaches the aquifer. We shall try two values of longitudinal dispersivity: 10 and 50 m.

In this case a is 50 m, $\alpha = \lambda = 0$, and longitudinal coefficients of dispersion are 1 and 5 m²/d, respectively. Appendix C includes the variation of concentration C/C_o as a function of x and y for v = 0.1 m/d, D_L = 1, D_T = 0.1 m²/d, and t = 1 and 5 years. Similar results for D_L = 5, D_T = 0.5 m²/d are also given there. Figures 9 and 10 show the variation of dimensionless concentration versus y for a longitudinal dispersion coefficient of 1 m²/d for values of x = 25, 50, 75, and 100 m, for t = 1 and 5 years, respectively. Similar results for a longitudinal dispersion coefficient of 5 m²/d are illustrated in Figures 11 and 12.

To complete the second part of this example, note that the dimensionless concentration C/C_o at the boundary of our zone of interest is 10/1000 = 0.01. From

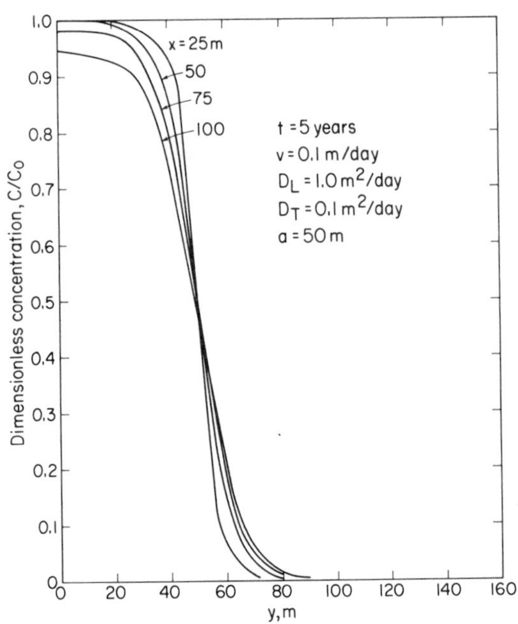

Fig. 10. Dimensionless concentration C/C_o versus distance from the x axis for values of x = 25, 50, 75, and 100 m downstream from the source and t = 5 years, v = 0.1 m/d, D_L = 1 m²/d, D_T = 0.1 m²/d, a = 50 m.

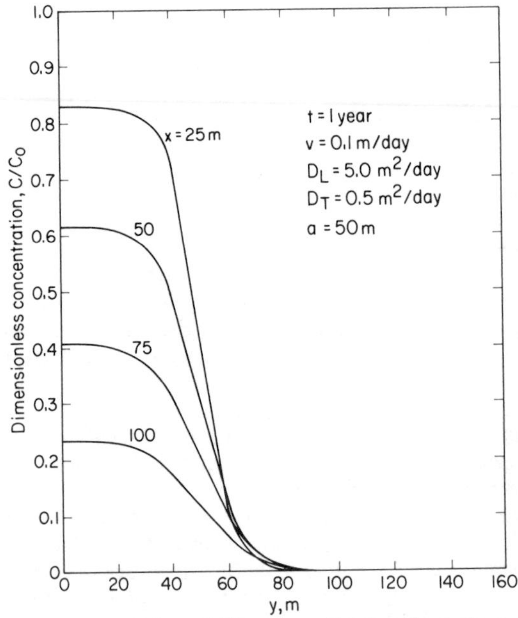

Fig. 11. Dimensionless concentration C/C_o versus distance from the x axis for values of x = 25, 50, 75, and 100 m downstream from the source and t = 1 year, v = 0.1 m/d, D_L = 5 m²/d, D_T = 0.5 m²/d, a = 50 m.

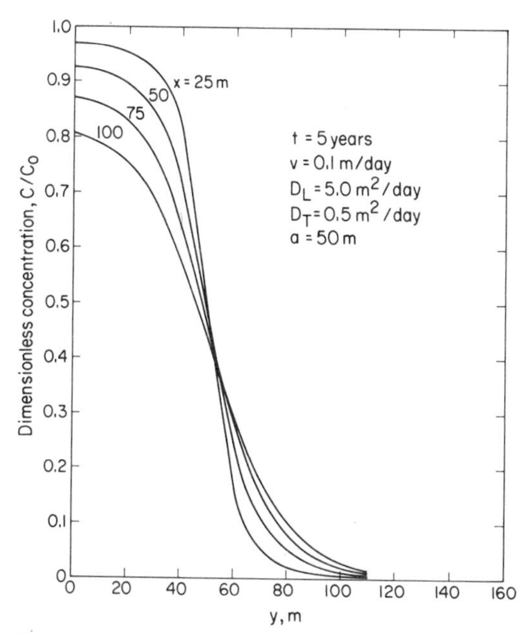

Fig. 12. Dimensionless concentration C/C_o versus distance from the x axis for values of x = 25, 50, 75, and 100 m downstream from the source and t = 5 years, v = 0.1 m/d, D_L = 5 m²/d, D_T = 0.5 m²/d, a = 50 m.

the Appendix C table corresponding to v = 0.1 m/d, D_L = 1, D_T = 0.1 m²/d, and t = 1825 days (5 years) locate the points with values of C/C_o just above and below 0.01. Figure 13 has been prepared by transferring these points to a map; the contour of C/C_o = 0.01 was constructed by linear interpolation. The contour in Figure 13 delineates that part of the aquifer with solute concentrations greater than 10 ppm 5 years after the solute reaches the aquifer. When the longitudinal dispersion coefficient D_L is 5 m²/d, the appropriate table from Appendix C should be used to construct the corresponding map in a similar fashion.

2.4.3. *Example 3*

This example illustrates the application of the approximate formula (60) given for advection-dispersion in radial flow and is useful for the examination of problems typically encountered in underground injection and pumpage to remove contaminants. A partially treated liquid waste is being continuously recharged, at a rate of 20 m³/h, into a well fully penetrating a homogeneous, isotropic confined aquifer of infinite horizontal extent, having a thickness of 10 m. The regional groundwater velocity in the vicinity of the well is negligible in comparison to the velocity generated by the recharge water. If the concentration of a nonreactive constituent in the recharging water is 2000 ppm, we will estimate the concentration in the groundwater at distances of 100, 500, and 1000 m away from the well after 10 years of operation. We will consider three different values for longitudinal dispersivity: α_L = 0.1, 1, and 10 m. A porosity of about 0.2 has been estimated in the aquifer.

In order to use (60), we first calculate the dimensionless parameters t_D and r_D for the three values of α_L (cases 1, 2, and 3).

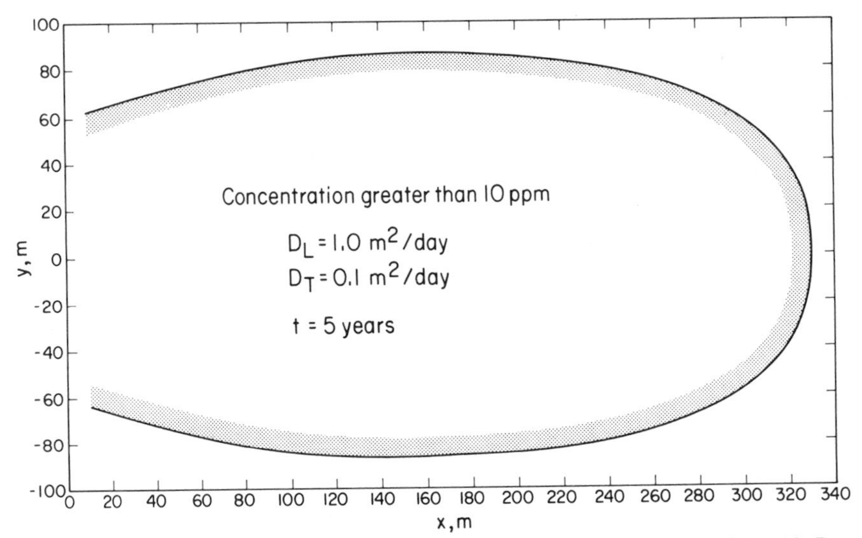

Fig. 13. A map showing the zone of contamination for example 2 where $v = 0.1$ m/d, $D_L = 1.0$ m²/d, $D_T = 0.1$ m²/d, and a time of 5 years after the solute reached the aquifer.

Case 1. Minimal dispersion: $\alpha_L = 0.1$ *m.*

$$t_D = \frac{Qt}{2\pi bn\alpha_L^2}$$

where

- Q rate of recharge, 20 m³/h;
- t 10 years or 87,600 hours;
- b thickness of aquifer, 10 m;
- n porosity, 0.2;
- α_L longitudinal dispersivity, 0.1 m.

$$t_D = \frac{(20)(87{,}600)}{(2\pi)(10)(0.2)(0.1)^2} = 13{,}941{,}973$$

(Note that any compatible system of units may be used.)

$$r_D = r/\alpha_L$$

where r is the radial distance from the recharging well, 100, 500, and 1000 m.

Thus the corresponding r_D's are 1000, 5000, and 10,000, respectively. Substituting the values of r_D and t_D into (60) we get

TABLE 6. Values of concentration in ppm at different distances from the well after 10 years of operation

r, m	α_L, m 0.1	1.0	10.0
100	2000	2000	2000
500	2000	1886	1383
550	0.382	261	722
600	0.0	0.720	285
1000	0.0	0.0	0.010

$$\frac{C}{C_o} = \frac{1}{2}\,\text{erfc}\left[\frac{\left(r_D^2/2\right) - t_D}{\left(4r_D^3/3\right)^{1/2}}\right] = \frac{1}{2}\,\text{erfc}\left[\frac{\left((1000)^2/2\right) - 13{,}941{,}973}{\left(4(1000)^3/3\right)^{1/2}}\right]$$

$$\frac{C}{C_o} = \frac{1}{2}\,\text{erfc}\,(-368) = \frac{1}{2}\,[1 + \text{erf}\,(368)]$$

From Appendix G, erf (368) = 1; thus $C/C_o = 1$, which indicates that the concentration at a distance r = 100 m after 10 years is equal to the recharge water concentration of 2000 ppm. Following the above procedure, one can easily determine values of concentration at 500 and 1000 m after 10 years. They are 2000 and 0.0 ppm, respectively.

Case 2. Mild dispersion: $\alpha_L = 1$ *m.* For this case t_D = 139,419 and values of r_D become 100, 500, and 1000. The ratio of C/C_o for this t_D and three values of r_D can be calculated following the above approach as 1, 0.943, and 0.0 for r = 100, 500, and 1000 m, respectively. This corresponds to concentrations of 2000, 1886, and 0.0 ppm at those distances.

Case 3. Moderate dispersion: $\alpha_L = 10$ *m.* In this case t_D = 1394 and values of r_D become 10, 50, and 100. The ratios of C/C_o for this case are 1., 0.69, and 0.5×10^{-5}, respectively, which correspond to concentrations of 2000, 1383, and 0.010 ppm at those distances.

The above calculations are summarized in Table 6. Additional data for other distances are also included. This data is plotted in Figure 14. From Table 6 and Figure 14 we see that for dispersivity values of 0.1 and 1 m, solute concentration at a distance of 1000 m is negligible. However, for α_L = 10 m, we begin to observe traces of the solute at a distance of 1000 m.

This example shows that, although the method is an approximation, in the absence of sufficient data one can easily determine the order of magnitude of the extent of a given solute in the aquifer without the use of a computer. For this example, it is obvious that the first 500-m interval from the recharge well is exposed to the given solute regardless of the value of longitudinal dispersivity α_L. The interval between r = 500 and 1000 m is the transition zone, and the zone beyond r = 1000 m is not exposed to the solute unless α_L is greater than 10 m.

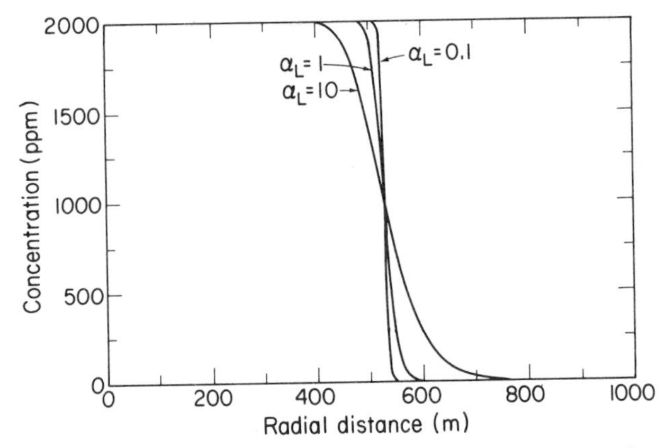

Fig. 14. Concentration versus radial distance for three different values of α_L based on the approximate formula (60) for advection-dispersion in radial flow.

2.5. Advantages of Analytical Methods

Some of the advantages of using analytical methods for estimating the extent of contamination in groundwater aquifers are listed below.

1. Analytical methods are probably the most efficient alternative when data necessary for identification of the system are sparse and uncertain.
2. Where applicable, these methods are the most economical approach.
3. They are always the most useful means for an initial estimation of the order of magnitude of contamination extent.
4. Experienced modelers and complex numerical codes are not required.
5. In many cases a rough estimate can be obtained through the tables supplied in the appendices of this report or other available sources. When application of simple computer codes for evaluation of analytical solutions is needed, the input data are usually very simple and do not require a detailed familiarity with the codes.

2.6. Limitations of Analytical Methods

Some of the important limitations of analytical methods are as follows.

1. The analytical solutions available are limited to certain idealized conditions and may not be applicable to a field problem with complex boundary conditions.
2. Spatial or temporal variation of system properties such as permeability and dispersivity cannot be handled with analytical techniques.

3 Semianalytical Methods

We shall now describe approximate techniques which in some respects are more powerful than analytical methods. Their application is much simpler than most of the complete numerical methods. These techniques apply a well-known concept of fluid mechanics, the complex velocity potential, and extend it by employing numerical tools and computer plotting capabilities.

A major limitation of these techniques is that they are only applicable to the study of steady state two-dimensional fluid flow through homogeneous media. Furthermore, the effects of transportation by dispersion and diffusion are not considered; contaminant species either move with the water—"water-coincident contaminants" or lag behind it due to adsorption on the rock matrix—"adsorption-retarded contaminants", as described in section 2.1.

In summary, execution of the semianalytical methods to determine contamination extent includes the following steps.

1. Identify simple flow components of the system such as uniform regional flow, point sources representing recharging wells, point sinks representing discharging wells, and finite radius circular sources representing waste storage ponds.
2. Combine the expressions for each of the identified simple flow components to obtain the overall complex velocity potential of the system, satisfying the appropriate boundary conditions.
3. Construct the expressions for the velocity potential and stream function of the system.
4. Calculate the velocity field by taking the derivative of the velocity potential.
5. Construct flow patterns and identify locations of any contaminant fronts for various values of time.
6. Using the stream function of the system, calculate the time variation of the rate at which a contaminant reaches any desired outflow boundary.

3.1. Theory

The theory behind the semianalytical methods can be summarized as follows. The analytic function $W = \phi + i\psi$ with the following properties is called the complex velocity potential.

1. Both ϕ and ψ are harmonic functions in that they satisfy Laplace's equation. In other words, $\nabla^2\phi = \nabla^2\psi = 0$. So defined, ϕ and ψ are conjugate harmonic functions.
2. The functions ϕ and ψ are the velocity potential and stream function, respectively.
3. Curves of velocity potentials ϕ = constant and streamlines ψ = constant intersect each other at right angles.

Velocity potential is generally defined as

$$\phi = Kh + c \tag{61}$$

Therefore a component of the specific discharge or Darcy velocity vector in any arbitrary direction x is

$$q_x = -\frac{\partial \phi}{\partial x} = -K\frac{\partial h}{\partial x} \tag{62}$$

The stream function of a flow system with a known velocity potential can be obtained simply by using the Cauchy-Riemann equations, which hold because of the properties of ϕ and ψ:

$$\frac{\partial \phi}{\partial x} = \frac{\partial \psi}{\partial y} \tag{63}$$

$$\frac{\partial \phi}{\partial y} = -\frac{\partial \psi}{\partial x} \tag{64}$$

It is important to note that the above theory is restricted to steady state, two-dimensional plane flow fields. Let us now review some basic, simple flow components which are of interest.

3.1.1. *Uniform Flow*

The complex velocity potential of a uniform flow with Darcy velocity U in a direction making an angle α with the positive x axis is given by

$$W = -UZe^{-i\alpha} + c \tag{65}$$

Substituting for complex numbers Z and $e^{-i\alpha}$, we get the velocity potential and the stream function for such a flow system.

$$W = \phi + i\psi = -UZe^{-i\alpha} + c = -U(x + iy)(cos\ \alpha - i \sin \alpha) + c \tag{66}$$

$$\phi = -U(x \cos \alpha + y \sin \alpha) + c_1 \tag{67}$$

$$\psi = U(x \sin \alpha - y \cos \alpha) + c_2 \tag{68}$$

Thus the following equations

$$\phi = -U(x \cos \alpha + y \cos \alpha) = constant \tag{69}$$

$$\psi = U(x \sin \alpha - y \cos \alpha) = constant \tag{70}$$

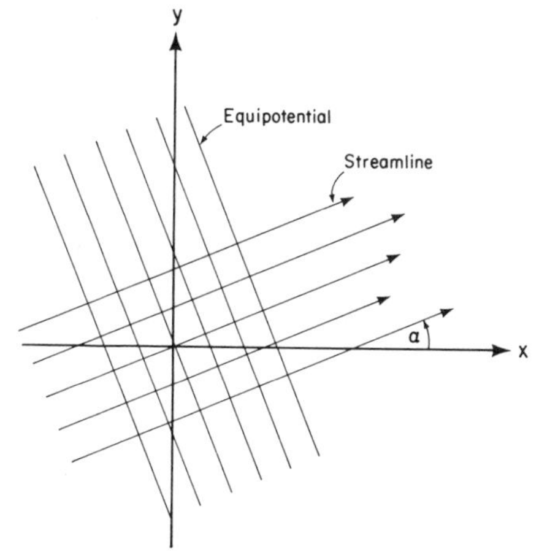

Fig. 15. A schematic diagram showing equipotentials and streamlines for a uniform flow making an angle α with the positive x axis.

represent equations for equipotentials and streamlines, respectively. Hence equipotentials are straight lines making an angle of $(\pi/2 + \alpha)$ with the positive x axis and streamlines are straight lines making an angle of α with the positive x axis (Figure 15). Components of specific discharge based on the above definitions are

$$q_x = -\frac{\partial \phi}{\partial x} = U \cos \alpha \tag{71}$$

$$q_y = -\frac{\partial \phi}{\partial y} = U \sin \alpha \tag{72}$$

3.1.2. *Sources and Sinks*

The complex velocity potential of a source with strength m located at the point Z_o is given by

$$W = m \, \ell n \, (Z - Z_o) + c \tag{73}$$

If the source represents a well which is being recharged at the rate of Q into an aquifer of thickness b, then the strength m of the source may be written as

$$m = -Q/2\pi b \tag{74}$$

Substituting for complex numbers Z and Z_o in (73), the velocity potential and stream function for such a flow system may be developed as follows:

$$W = \phi + i\psi = \frac{-Q}{2\pi b}\,\ell n(Z - Z_o) + c$$

$$= \frac{-Q}{2\pi b}\,\ell n[(x - x_o)^2 + (y - y_o)^2]^{1/2} - i\frac{Q}{2\pi b}\tan^{-1}\left[\frac{y - y_o}{x - x_o}\right] + c \qquad (75)$$

$$\phi = \frac{-Q}{4\pi b}\,\ell n[(x - x_o)^2 + (y - y_o)^2] + c_1 \qquad (76)$$

$$\psi = \frac{-Q}{2\pi b}\tan^{-1}\left[\frac{y - y_o}{x - x_o}\right] + c_2 \qquad (77)$$

where x_o and y_o are the coordinates of the source and x and y are the coordinates of a point where ϕ and ψ are calculated. Thus the following equations

$$\phi = \frac{-Q}{4\pi b}\,\ell n[(x - x_o)^2 + (y - y_o)^2] = constant \qquad (78)$$

$$\psi = \frac{-Q}{2\pi b}\tan^{-1}\left[\frac{y - y_o}{x - x_o}\right] = constant \qquad (79)$$

are the equations for equipotentials and streamlines, respectively. Obviously, equipotentials are concentric circles with the center at the source. Streamlines are a family of radial lines emanating from the source. Components of specific discharge based on the above definitions are

$$q_x = -\frac{\partial \phi}{\partial x} = \frac{Q}{2\pi b}\,\frac{(x - x_o)}{(x - x_o)^2 + (y - y_o)^2} \qquad (80)$$

$$q_y = -\frac{\partial \phi}{\partial y} = \frac{Q}{2\pi b}\,\frac{(y - y_o)}{(x - x_o)^2 + (y - y_o)^2} \qquad (81)$$

The above describes a source from which the flow diverges. For a sink to which the flow converges, the strength m is positive.

3.1.3. *Doublet*

Let us assume a source and sink of equal strength m, located at equal distance d from the origin, in opposite directions along the x axis. Now, if we let the source and the sink come together, i.e., let d tend to zero and at the same time increase the strength m to infinite value such that in the limit the product of dm/π is some finite number Ω, the combination of this sink and source is called a doublet. A doublet is considered to be positive when flow proceeds out from the origin in the positive x direction and returns to the origin from the negative x direction.

The complex velocity potential for a positive doublet is

$$W = \frac{\Omega \bar{Z}}{|Z|^2} + c \tag{82}$$

where $\bar{Z}$ and $|Z|$ are conjugate and modulus of complex number Z, respectively.

Substituting for $\bar{Z}$ and $|Z|$, we get the velocity potential and stream function for a positive doublet located at the origin:

$$W = \phi + i\psi = \frac{\Omega(x - iy)}{x^2 + y^2} + c \tag{83}$$

$$\phi = \frac{\Omega x}{x^2 + y^2} + c_1 \tag{84}$$

$$\psi = -\frac{\Omega y}{x^2 + y^2} + c_2 \tag{85}$$

Therefore equations for equipotentials and streamlines may be written as

$$\phi = \frac{\Omega x}{x^2 + y^2} = \mathit{constant} \tag{86}$$

$$\psi = \frac{-\Omega y}{x^2 + y^2} = \mathit{constant} \tag{87}$$

Equations (86) and (87) may be rearranged as

$$\left(x - \frac{\Omega}{2c}\right)^2 + y^2 = \left(\frac{\Omega}{2c}\right)^2 \tag{88}$$

$$x^2 + \left(y + \frac{\Omega}{2c}\right)^2 = \left(\frac{\Omega}{2c}\right)^2 \tag{89}$$

Equation (88) represents equipotentials and describes circles with centers along the x axis. The radius and center of each circle correspond to the value of c, but all of them pass through the origin and thus are tangent to the y axis. Similarly, (89) represents a group of circles with centers on the y axis and tangent to the x axis. These circles are the streamlines for such a doublet.

3.2. Combination of Uniform Flow With Point Sources and Sinks

Due to the linearity of Laplace's equation, one can superimpose as many flow components as required to obtain the expression for the complex velocity potential of an entire system. In this section we shall concentrate on problems consisting of one or several point sources of contaminant recharge, together with some groundwater discharging wells, combined with a uniform regional groundwater flow regime. The overall complex velocity potential of such a system may be written as

$$W = -UZe^{-i\alpha} + \sum_{j=1}^{N} \frac{Q_j}{2\pi b} \ell n(Z - Z_j) - \sum_{k=1}^{M} \frac{Q_k}{2\pi b} \ell n(Z - Z_k) + c \tag{90}$$

where

- W overall complex velocity potential of the system;
- U Darcy velocity of uniform regional flow;
- α the angle between the direction of regional flow and the positive x axis;
- b aquifer thickness;
- Q_j rate of discharge from well j;
- Q_k rate of recharge of well k.

The first term on the right-hand side of (90) represents uniform regional flow. The second term is for various discharging wells j, representing sinks, located at the points Z_j on the complex plane. The third term is for recharging wells k, representing sources, located at $Z = Z_k$. The constant c is determined so that W satisfies the boundary conditions of the problem. N and M are the number of discharging and recharging wells, respectively.

Once the complex velocity potential of the system is determined, one can easily find the expressions for velocity potential and stream function of the combined simple flow components. The expression for the velocity potential of the above system, the real part of W, is

$$\phi = -U(x \cos\alpha + y \sin\alpha) + \sum_{j=1}^{N} \frac{Q_j}{4\pi b} \ell n\,[(x - x_j)^2 + (y - y_j)^2]$$

$$- \sum_{k=1}^{M} \frac{Q_k}{4\pi b} \ell n\,[(x - x_k)^2 + (y - y_k)^2] + c_1 \tag{91}$$

and the expression for the stream function, the imaginary part of W, becomes

$$\psi = U(x \sin \alpha - y \cos \alpha) + \sum_{j=1}^{N} \frac{Q_j}{2\pi b} \tan^{-1}\left[\frac{y - y_j}{x - x_j}\right]$$

$$- \sum_{k=1}^{M} \frac{Q_k}{2\pi b} \tan^{-1}\left[\frac{y - y_k}{x - x_k}\right] + c_2 \tag{92}$$

The velocity field of the system may be easily established by using (62). At any given point with coordinate (x, y), components of the specific discharge for the overall system may be written as

$$q_x = -\frac{\partial \phi}{\partial x} = U \cos \alpha - \sum_{j=1}^{N} \frac{Q_j}{2\pi b} \frac{(x - x_j)}{(x - x_j)^2 + (y - y_j)^2}$$

$$+ \sum_{k=1}^{M} \frac{Q_k}{2\pi b} \frac{(x - x_k)}{(x - x_k)^2 + (y - y_k)^2} \tag{93}$$

$$q_y = -\frac{\partial \phi}{\partial y} = U \sin \alpha - \sum_{j=1}^{N} \frac{Q_j}{2\pi b} \frac{(y - y_j)}{(x - x_j)^2 + (y - y_j)^2}$$

$$+ \sum_{k=1}^{M} \frac{Q_k}{2\pi b} \frac{(y - y_k)}{(x - x_k)^2 + (y - y_k)^2} \tag{94}$$

Components of the average pore water velocity for an individual fluid particle moving through the overall flow system may be written as

$$v_x = q_x / n \qquad v_y = q_y / n \tag{95}$$

where n is porosity of the medium and q_x and q_y are components of specific discharge which were given by (93) and (94), respectively.

Because of adsorption onto the rock matrix, the velocity at which a contaminant species flows through the aquifer may be less than the pore water velocity. As was discussed earlier, for some simplified cases this is accounted for by introducing a retardation factor R, which was defined by (7). The retardation factor R is the ratio of pore water velocity to the velocity of that particular chemical. In this case, components of contaminant velocity become

$$v_{cx} = q_x / nR \qquad v_{cy} = q_y / nR \tag{96}$$

When $R = 1$ there is no adsorption and the contaminant velocity is identical to the pore water velocity. Such a contaminant species is called a "water-coincident contaminant." When $R > 1$ the contaminant species is called an "adsorption-retarded contaminant."

The path line traveled by a contaminant particle can be divided into increments $d\ell$, which are traversed in time intervals dt. The projections of $d\ell$ on the x and y axes are given by dx and dy, respectively, where

$$dx = v_{cx}\, dt = q_x dt / nR \tag{97}$$

$$dy = v_{cy}\, dt = q_y dt / nR \tag{98}$$

and

$$d\ell = (dx^2 + dy^2)^{1/2} = (q_x^2 + q_y^2)^{1/2}\, dt / nR \tag{99}$$

Numerical integration of (99) yields travel time between any two points of a given streamline. Furthermore, if a contaminant particle is at a point (x_j, y_j) at time t, its new position at time $t + \Delta t$ on the same streamline can be calculated by use of the following equations:

$$x_{j+1} = x_j + \Delta x = x_j + q_x \Delta t / nR \tag{100}$$

$$y_{j+1} = y_j + \Delta y = y_j + q_y \Delta t / nR \tag{101}$$

Thus one can easily follow the path of each particle by using the above equations.

A computer code called RESSQ can be used to perform the above calculation numerically. RESSQ draws the flow pattern in the aquifer by tracing streamlines from injection wells. Since the length of the streamlines connecting any pair of recharge and discharge wells is variable, the required time for transport of contaminant from the recharge well to the producing well is different along various streamlines. RESSQ calculates the concentration of a given solute at a production well as a function of time based on the arrival of streamlines. This code is also capable of calculating the location of the water-coincident or adsorption-retarded contaminant front around an injection well at any given time. A listing of the code RESSQ together with input descriptions is given in Appendix H. Examples of its use are given in section 3.5.

3.3. Combination of Uniform Flow With a Finite Radius Source

Sometimes when a source of contamination covers a large area, simulating it with a point source may introduce a significant error into the final results. In order to avoid such errors a method introduced by *Nelson* [1978] is employed. The method accounts for a circular source of finite radius discharging contaminated fluid into the aquifer under a constant head.

It can be shown that the combination of a uniform flow in the positive x direction with a positive doublet and a point source both centered at the origin represents outflow from a completely penetrating cylindrical pond in the presence of a uniform flow in the positive x direction. The complex velocity potential for such a combination may be written as

$$W = -UZ + \frac{\Omega \bar{Z}}{|Z|^2} - \frac{Q_p}{2\pi b} \ell n\ Z + c \tag{102}$$

The velocity potential and stream function for such a flow system may be obtained from the real and imaginary part of W:

$$\phi = -Ux + \frac{\Omega x}{x^2 + y^2} - \frac{Q_p}{4\pi b} \ell n(x^2 + y^2) + c_1 \tag{103}$$

$$\psi = -Uy - \frac{\Omega y}{x^2 + y^2} - \frac{Q_p}{2\pi b} \tan^{-1}\left(\frac{y}{x}\right) + c_2 \tag{104}$$

where

U Darcy velocity of uniform flow in the positive x direction;
Q_p rate of outflow from the pond;
b thickness of the aquifer;
Ω constant of the doublet.

The value of the constants c_1 and Ω in (103) can be determined such that ϕ satisfies applicable boundary conditions. If we hold ϕ constant and equal to H_0 at $r = r_0$, (103) may be written as

$$\phi = H_0 - Ux + \frac{U r_0^2 x}{x^2 + y^2} - \frac{Q_p}{4\pi b} \ell n \left(\frac{x^2 + y^2}{r_0^2}\right) \tag{105}$$

If some sources and sinks are also present in the field, one can easily incorporate the velocity potential of those with (105). The result is

$$\phi = H_0 - Ux + \frac{U r_0^2 x}{x^2 + y^2} - \frac{Q_p}{4\pi b} \ell n \left(\frac{x^2 + y^2}{r_0^2}\right)$$

$$+ \sum_{j=1}^{N} \frac{Q_j}{4\pi b} \ell n \left(\frac{(x - x_j)^2 + (y - y_j)^2}{x_j^2 + y_j^2}\right)$$

$$- \sum_{k=1}^{M} \frac{Q_k}{4\pi b} \ell n \left[\frac{(x - x_k)^2 + (y - y_k)^2}{x_k^2 + y_k^2} \right] \tag{106}$$

where Q_j and Q_k are rates of discharge and recharge of sinks and sources, respectively. Note that the arguments of logarithms in (106) have been modified such that the effect of sources and sinks vanish at the origin. This is an approximation to the requirement that they vanish all over the circle with radius r_0 centered at the origin. This approximation is valid when $x_j^2 + y_j^2$ is greater than r_0^2. Equation (106) is defined in the region outside the pond $x^2 + y^2 > r_0^2$, except at the points with coordinates x_j, y_j and x_k, y_k. Components of the average pore water velocity at any point (x, y) within the overall flow system where ϕ is defined may be written as

$$v_x = -\frac{1}{n}\frac{\partial \phi}{\partial x} = \frac{U}{n} + \frac{U r_0^2}{n} \left[\frac{x^2 - y^2}{(x^2 + y^2)^2} \right] + \frac{Q_p}{2\pi n b} \frac{x}{x^2 + y^2}$$

$$- \sum_{j=1}^{N} \frac{Q_j}{2\pi n b} \frac{x - x_j}{(x - x_j)^2 + (y - y_j)^2} + \sum_{k=1}^{M} \frac{Q_k}{2\pi n b} \frac{x - x_k}{(x - x_k)^2 + (y - y_k)^2} \tag{107}$$

$$v_y = -\frac{1}{n}\frac{\partial \phi}{\partial y} = \frac{U r_0^2}{n} \left[\frac{2xy}{(x^2 + y^2)^2} \right] + \frac{Q_p}{2\pi n b} \frac{y}{x^2 + y^2}$$

$$- \sum_{j=1}^{N} \frac{Q_j}{2\pi n b} \frac{y - y_j}{(x - x_j)^2 + (y - y_j)^2} + \sum_{k=1}^{M} \frac{Q_k}{2\pi n b} \frac{y - y_k}{(x - x_k)^2 + (y - y_k)^2} \tag{108}$$

With the components of pore water velocity thus determined, the approach explained in section 3.2 is used to calculate the travel time of a contaminant particle between a source of contamination and any other point in the system. RESSQ, the computer code supplied in Appendix H can be used to facilitate this task.

3.4. Use of a Single Producing Well for Monitoring Purposes

In this section we show how a single producing well can be used to map a contaminant concentration distribution within an aquifer. As we saw before, the components of the average pore water velocity due to a producing well located at (x_o, y_o), with a discharge rate of Q, are

$$v_x = \frac{-Q}{2\pi n b} \frac{(x - x_o)}{(x - x_o)^2 + (y - y_o)^2} \tag{109}$$

$$v_y = \frac{-Q}{2\pi nb} \frac{(y - y_o)}{(x - x_o)^2 + (y - y_o)^2} \tag{110}$$

In radial coordinates, the components of the average pore water velocity for a sink located at the origin are

$$v_r = -Q/2\pi nbr \tag{111}$$

$$v_\theta = 0 \tag{112}$$

Substituting dr/dt for v_r we get

$$v_r = \frac{dr}{dt} = \frac{-Q}{2\pi nbr} \tag{113}$$

or

$$\int_{t_1}^{t_2} dt = -\int_{r_1}^{r_2} \frac{2\pi nbrdr}{Q} \tag{114}$$

Integrating and solving for r_2 gives

$$r_2 = \left[r_1^2 - \frac{Q(t_2 - t_1)}{n\pi b} \right]^{1/2} \tag{115}$$

If an observation well is located at a distance r_1 from the pumping well, then a time series of contaminant concentration measurements at that well taken at times $t_1, t_1', t_1'', \cdots$ will yield the corresponding locations $r_2, r_2', r_2'', \cdots$ for those concentrations at any given time t_2. Hence, assuming that the concentration distribution of a given solute in an aquifer is not uniform, the time series data from a given well can be mapped out into the aquifer to produce a "snapshot" of the spatial contaminant concentration profile along the radial direction from the production well to the observation well at various times. By using observation wells in several directions from the pumping well, an areal picture of the contaminant concentration in the aquifer at various times can be determined. This technique of correlating time series data from a single well to a spatial distribution for a given time has been developed from the techniques described by *Keely* [1982].

A simple computer program called RT has been written to implement the above concept. A listing of this program and instructions for its use are given in Appendix I. An example of its use is given in section 3.5.

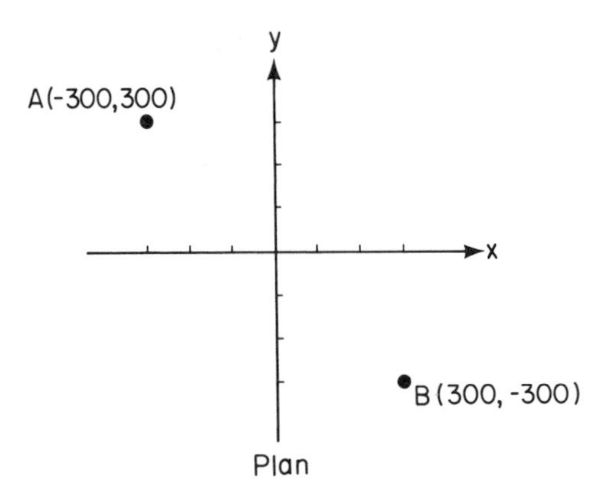

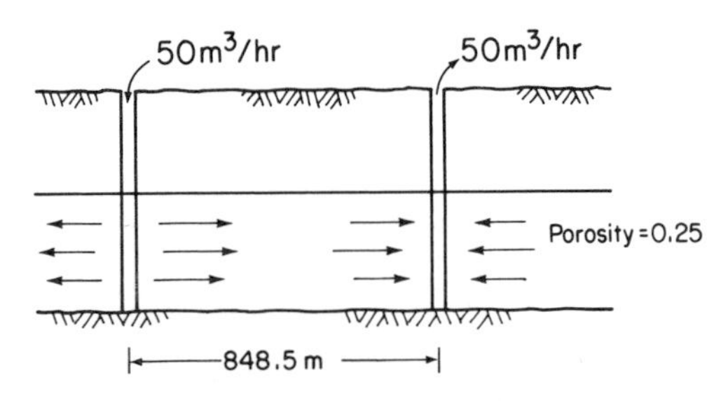

Fig. 16. A schematic diagram showing the plan and vertical section of the recharge and discharge wells used in examples 1 and 2.

3.5. Applications

Several examples have been designed to illustrate the application of the semi-analytical techniques. Emphasis has been placed on the application of the related computer code in the field. Note that due to the restriction of these solutions to the water-coincident or adsorption-retarded solute movement, the contribution of mass transport by dispersion and diffusion is not taken into account in the following examples.

3.5.1. *Example 1*

Let us consider a well located at point A of Figure 16, completely penetrating a homogeneous, isotropic aquifer with a thickness of 10 m. Partially treated waste-water is being recharged at a rate of 50 m^3/h into the aquifer through this well. Another well at point B, 848.5 m from well A, is being continuously pumped at a rate of 50 m^3/h from the same aquifer. Effective porosity of the aquifer in this region is known to be 25%; regional flow and adsorption are assumed negligible.

Given the above conditions, the following items are required.

1. Show the pattern of streamlines for the above flow system.
2. Draw the position of the water-coincident contaminant in the aquifer after 0.5, 2, and 4 years.

TABLE 7. Input data for program RESSQ for example 1

Card	Columns	Entry	Description
1	1−80	Example one	Title
2	1−5	1	Number of injection wells
	6−10	1	Number of production wells
	11−20	0.	Ambient concentration in aquifer
	21−30	100.	Injection concentration
	31−40		Blank (default) for using concentration in percentage
	41−45	2	For practical units
3	1−10	10.	Thickness of aquifer, m
	11−20	25.	Porosity, percent
	21−30	0.	Pore velocity of uniform regional flow, m/yr
	31−40	0.	Direction of uniform flow, degrees
	41−50	0.	No adsorption (R=1 in (96))
4	1−10	Injwell−1	Name of the injection well
	11−20	−300	x coordinate, m
	21−30	300.	y coordinate, m
	31−40	50.	Rate of injection, m^3/h
	41−50		Blank means use default well radius r_w = 7.5 cm
	51−60	100.	Injection concentration
	61−70	0.	Angle at which the first streamline calculated leaves the well
	71−72	45	Number of streamlines to be calculated
	73−74	3	Every 3 streamlines to be plotted
	75−76		Blank means fronts are to be plotted around this well
5	1−10	Prodwell−1	Name of the production well
	11−20	300.	x coordinate, m
	21−30	−300.	y coordinate, m
	31−40	50.	Rate of production, m^3/h
	41−50		Blank means use default wall radius r_w = 7.5 cm
	51−60		Blank means production concentration is to be studied
6	1−10	3	Number of fronts to be calculated
	11−20	0.5	Time of the first front, years
	21−30	2.	Time of the second front, years
	31−40	4.	Time of the third front, years
7	1−10	200.	Maximum amount of time for calculating the trace of a streamline, years
	11−20	10.	Step length for tracing out streamlines, m
	21−25		Blank to plot streamlines
	26−30		Blank to plot fronts
8	1−10	−1000.	Minimum x for plot, m
	11−20	1000.	Maximum x for plot, m
	21−30	−1000.	Minimum y for plot, m
	31−40	1000.	Maximum y for plot, m

3. Determine how long it will take for a water-coincident contaminant to reach production well B.
4. Determine the time variation of contaminant concentration at the producing well.

Before solving the above problem by the semianalytical technique, it is interesting to note that *Muskat* [1937] gave an analytical solution for calculating the front position in this problem at different times. He also gave the following formula which enables one to calculate the time required for the injected water to reach the pumping well:

$$t = \frac{4\pi n d^2 b}{3Q} \tag{116}$$

where d is one half of the distance between two wells. Based on (116), the required travel time for a particle of fluid to move from well A to well B is about 4.3 years.

The input data needed to run the program RESSQ given in Appendix H is shown in Table 7. Samples of the output produced by the program RESSQ for the above input data are shown in Tables 8 through 11. Table 8 shows the arrival time of each streamline at the production well. Note that nine of the streamlines have not reached the production well during the specified period of study, 200 years. The angle at which each streamline leaves the recharging well is also shown on this table. Streamlines 40 and 41, leaving the recharge well at angles of 312 and 320, respectively, are the first ones to arrive at the production well. They arrive 4.3 years after they leave the recharge well, which is exactly what we calculated earlier from (116).

Table 9 presents the time variation of contaminant concentration at the production well. Note that a total of 45 stream tubes emanate from the injection well. Thus each of these stream tubes carries 1/45 of the total injection rate, which in this example is equal to the production rate of 50 m^3/h. Hence the mixing of wastewater from each of these stream tubes with fresh water at the production well increases the contaminant concentration of the produced water by 1/45 = 0.0222. Therefore the arrival of a total of 36 streamtubes at the production well after 138 years has increased the concentration by 80% (36 $\times$ 0.0222 = 0.80). If the injected fluid contains 100 ppm of contaminant, the production well draws in 80 ppm. No other streamline arrives at the production well in the time interval between 138 and 200 years, so the contaminant concentration of the producted water will not increase by more than 2.2% during that time. Figure 17 illustrates the time variation of contaminant concentration at the production well.

Table 10 shows the coordinates (x, y) of the points along some of the streamlines. Coordinate lists are printed for every streamline to be plotted, as specified in the input data. Figure 18, which has been prepared from data such as that given in Table 10, illustrates the pattern of streamlines between the injection and production well.

Table 11 gives the coordinates of the points along each of the water-coincident contaminant fronts at the particular times specified in the input data. A plot of the position of these fronts is also shown in Figure 18.

3.5.2. *Example 2*

For this example, we will again consider the configuration used in example 1, but we will add a uniform regional flow with a pore water velocity of 50 m/yr

TABLE 8. RESSQ output: arrival time of streamlines at the production well

STREAMLINES DEPARTING FROM INJECTION WELL INJWELL−1

NUMBER OF STREAMLINE	WELL REACHED	TIME OF ARRIVAL	ANGLE BETA IN DEGREES
1	PRODWELL−1	5.6 YEARS	0.0
2	PRODWELL−1	6.2 YEARS	8.0
3	PRODWELL−1	7.0 YEARS	16.0
4	PRODWELL−1	8.1 YEARS	24.0
5	PRODWELL−1	9.6 YEARS	32.0
6	PRODWELL−1	11.6 YEARS	40.0
7	PRODWELL−1	14.4 YEARS	48.0
8	PRODWELL−1	18.6 YEARS	56.0
9	PRODWELL−1	24.8 YEARS	64.0
10	PRODWELL−1	34.5 YEARS	72.0
11	PRODWELL−1	50.7 YEARS	80.0
12	PRODWELL−1	79.9 YEARS	88.0
13	PRODWELL−1	138.0 YEARS	96.0
14	+++NONE+++	200.9 YEARS	104.0
15	+++NONE+++	201.1 YEARS	112.0
16	+++NONE+++	200.9 YEARS	120.0
17	+++NONE+++	201.4 YEARS	128.0
18	+++NONE+++	201.6 YEARS	136.0
19	+++NONE+++	200.4 YEARS	144.0
20	+++NONE+++	200.3 YEARS	152.0
21	+++NONE+++	200.8 YEARS	160.0
22	+++NONE+++	200.0 YEARS	168.0
23	PRODWELL−1	119.1 YEARS	176.0
24	PRODWELL−1	70.7 YEARS	184.0
25	PRODWELL−1	45.8 YEARS	192.0
26	PRODWELL−1	31.6 YEARS	200.0
27	PRODWELL−1	23.0 YEARS	208.0
28	PRODWELL−1	17.4 YEARS	216.0
29	PRODWELL−1	13.6 YEARS	224.0
30	PRODWELL−1	11.0 YEARS	232.0
31	PRODWELL−1	9.2 YEARS	240.0
32	PRODWELL−1	7.8 YEARS	248.0
33	PRODWELL−1	6.8 YEARS	256.0
34	PRODWELL−1	6.0 YEARS	264.0
35	PRODWELL−1	5.5 YEARS	272.0
36	PRODWELL−1	5.0 YEARS	280.0
37	PRODWELL−1	4.7 YEARS	288.0
38	PRODWELL−1	4.5 YEARS	296.0
39	PRODWELL−1	4.4 YEARS	304.0
40	PRODWELL−1	4.3 YEARS	312.0
41	PRODWELL−1	4.3 YEARS	320.0
42	PRODWELL−1	4.4 YEARS	328.0
43	PRODWELL−1	4.5 YEARS	336.0
44	PRODWELL−1	4.8 YEARS	344.0
45	PRODWELL−1	5.1 YEARS	352.0

TABLE 9. RESSQ output: time variation of concentration at the production well

EVOLUTION OF CONCENTRATION FOR PRODUCTION WELL PRODWELL−1

TIME IN YEARS	CONCENTRATION IN PERCENT	(C−C0)/(CD−C0)
4.304	2.222E+00	0.0222
4.313	4.444E+00	0.0444
4.365	6.667E+00	0.0667
4.391	8.889E+00	0.0889
4.500	1.111E+01	0.1111
4.546	1.333E+01	0.1333
4.716	1.556E+01	0.1556
4.784	1.778E+01	0.1778
5.027	2.000E+01	0.2000
5.122	2.222E+01	0.2222
5.453	2.444E+01	0.2444
5.581	2.667E+01	0.2667
6.024	2.889E+01	0.2889
6.194	3.111E+01	0.3111
6.784	3.333E+01	0.3333
7.010	3.556E+01	0.3556
7.796	3.778E+01	0.3778
8.100	4.000E+01	0.4000
9.162	4.222E+01	0.4222
9.573	4.444E+01	0.4444
11.027	4.667E+01	0.4667
11.598	4.889E+01	0.4889
13.636	5.111E+01	0.5111
14.445	5.333E+01	0.5333
17.383	5.556E+01	0.5556
18.568	5.778E+01	0.5778
22.951	6.000E+01	0.6000
24.757	6.222E+01	0.6222
31.598	6.444E+01	0.6444
34.486	6.667E+01	0.6667
45.778	6.889E+01	0.6889
50.699	7.111E+01	0.7111
70.743	7.333E+01	0.7333
79.850	7.556E+01	0.7556
119.079	7.778E+01	0.7778
137.961	8.000E+01	0.8000

TABLE 10. RESSQ output: list of coordinates of the points along three streamlines

POINTS ON STREAMLINE 1 FROM INJWELL−1 - ENDS IN PRODWELL−1
NUMBER OF POINTS= 33

X	Y	X	Y	X	Y	X	Y
−.300E+03	0.300E+03	−.280E+03	0.300E+03	−.250E+03	0.298E+03	−.220E+03	0.296E+03
−.190E+03	0.291E+03	−.161E+03	0.285E+03	−.132E+03	0.278E+03	−.103E+03	0.269E+03
−.751E+02	0.259E+03	−.475E+02	0.247E+03	−.205E+02	0.234E+03	0.577E+01	0.219E+03
0.313E+02	0.204E+03	0.560E+02	0.187E+03	0.798E+02	0.168E+03	0.103E+03	0.149E+03
0.124E+03	0.128E+03	0.145E+03	0.107E+03	0.165E+03	0.841E+02	0.183E+03	0.604E+02
0.201E+03	0.359E+02	0.217E+03	0.105E+02	0.231E+03	−.156E+02	0.245E+03	−.425E+02
0.257E+03	−.700E+02	0.267E+03	−.981E+02	0.276E+03	−.127E+03	0.284E+03	−.156E+03
0.290E+03	−.185E+03	0.295E+03	−.215E+03	0.298E+03	−.244E+03	0.300E+03	−.274E+03
0.300E+03	−.294E+03						

TABLE 10. (continued)

POINTS ON STREAMLINE 4 FROM INJWELL−1 - ENDS IN PRODWELL−1
NUMBER OF POINTS= 38

X	Y	X	Y	X	Y	X	Y
−.300E+03	0.300E+03	−.282E+03	0.308E+03	−.253E+03	0.319E+03	−.225E+03	0.327E+03
−.196E+03	0.334E+03	−.166E+03	0.339E+03	−.136E+03	0.341E+03	−.106E+03	0.342E+03
−.761E+02	0.341E+03	−.463E+02	0.338E+03	−.167E+02	0.332E+03	0.124E+02	0.325E+03
0.410E+02	0.316E+03	0.689E+02	0.305E+03	0.961E+02	0.293E+03	0.122E+03	0.278E+03
0.148E+03	0.262E+03	0.172E+03	0.244E+03	0.194E+03	0.224E+03	0.216E+03	0.204E+03
0.236E+03	0.181E+03	0.255E+03	0.158E+03	0.271E+03	0.133E+03	0.287E+03	0.107E+03
0.300E+03	0.803E+02	0.312E+03	0.526E+02	0.322E+03	0.243E+02	0.330E+03	−.462E+01
0.336E+03	−.340E+02	0.340E+03	−.637E+02	0.342E+03	−.937E+02	0.342E+03	−.124E+03
0.340E+03	−.154E+03	0.336E+03	−.183E+03	0.330E+03	−.213E+03	0.322E+03	−.242E+03
0.313E+03	−.270E+03	0.303E+03	−.293E+03				

POINTS ON STREAMLINE 16 FROM INJWELL−1 - ENDS IN +++NONE+++
NUMBER OF POINTS= 73

X	Y	X	Y	X	Y	X	Y
−.300E+03	0.300E+03	−.310E+03	0.317E+03	−.325E+03	0.344E+03	−.339E+03	0.370E+03
−.352E+03	0.397E+03	−.365E+03	0.424E+03	−.378E+03	0.451E+03	−.390E+03	0.478E+03
−.402E+03	0.506E+03	−.413E+03	0.534E+03	−.424E+03	0.562E+03	−.434E+03	0.590E+03
−.443E+03	0.619E+03	−.453E+03	0.647E+03	−.461E+03	0.676E+03	−.469E+03	0.705E+03
−.477E+03	0.734E+03	−.484E+03	0.763E+03	−.490E+03	0.792E+03	−.496E+03	0.822E+03
−.502E+03	0.851E+03	−.507E+03	0.881E+03	−.511E+03	0.910E+03	−.515E+03	0.940E+03
−.518E+03	0.970E+03	−.521E+03	0.100E+04	−.523E+03	0.103E+04	−.525E+03	0.106E+04
−.526E+03	0.109E+04	−.527E+03	0.112E+04	−.527E+03	0.115E+04	−.527E+03	0.118E+04
−.526E+03	0.121E+04	−.524E+03	0.124E+04	−.522E+03	0.127E+04	−.519E+03	0.130E+04
−.516E+03	0.133E+04	−.513E+03	0.136E+04	−.508E+03	0.139E+04	−.504E+03	0.142E+04
−.498E+03	0.145E+04	−.493E+03	0.148E+04	−.486E+03	0.151E+04	−.479E+03	0.154E+04
−.472E+03	0.156E+04	−.464E+03	0.159E+04	−.456E+03	0.162E+04	−.447E+03	0.165E+04
−.437E+03	0.168E+04	−.427E+03	0.171E+04	−.417E+03	0.174E+04	−.406E+03	0.176E+04
−.394E+03	0.179E+04	−.382E+03	0.182E+04	−.370E+03	0.185E+04	−.357E+03	0.187E+04
−.343E+03	0.190E+04	−.330E+03	0.193E+04	−.315E+03	0.195E+04	−.300E+03	0.198E+04
−.285E+03	0.201E+04	−.269E+03	0.203E+04	−.253E+03	0.206E+04	−.236E+03	0.208E+04
−.219E+03	0.211E+04	−.201E+03	0.213E+04	−.183E+03	0.215E+04	−.165E+03	0.218E+04
−.146E+03	0.220E+04	−.127E+03	0.222E+04	−.107E+03	0.225E+04	−.870E+02	0.227E+04
−.665E+02	0.229E+04						

TABLE 11. RESSQ output: list of coordinates of the points along each recharged−water front

LINES TO FORM THE 6.0000 MONTHS FRONT AROUND INJWELL−1
NUMBER OF POINTS= 46

X	Y	X	Y	X	Y	X	Y
−.126E+03	0.276E+03	−.126E+03	0.298E+03	−.128E+03	0.320E+03	−.134E+03	0.341E+03
−.142E+03	0.363E+03	−.154E+03	0.383E+03	−.168E+03	0.402E+03	−.185E+03	0.419E+03
−.205E+03	0.433E+03	−.227E+03	0.445E+03	−.250E+03	0.453E+03	−.275E+03	0.458E+03
−.300E+03	0.460E+03	−.325E+03	0.457E+03	−.350E+03	0.450E+03	−.373E+03	0.440E+03
−.395E+03	0.426E+03	−.414E+03	0.409E+03	−.430E+03	0.389E+03	−.443E+03	0.367E+03
−.452E+03	0.344E+03	−.458E+03	0.319E+03	−.460E+03	0.293E+03	−.457E+03	0.268E+03
−.451E+03	0.244E+03	−.442E+03	0.221E+03	−.430E+03	0.200E+03	−.415E+03	0.181E+03
−.397E+03	0.165E+03	−.378E+03	0.151E+03	−.357E+03	0.140E+03	−.336E+03	0.132E+03
−.314E+03	0.127E+03	−.292E+03	0.125E+03	−.271E+03	0.126E+03	−.250E+03	0.130E+03
−.230E+03	0.136E+03	−.211E+03	0.144E+03	−.194E+03	0.155E+03	−.178E+03	0.167E+03
−.164E+03	0.182E+03	−.152E+03	0.198E+03	−.142E+03	0.216E+03	−.134E+03	0.235E+03
−.129E+03	0.255E+03	−.126E+03	0.276E+03				

TABLE 11. (continued)

LINES TO FORM THE 2.0000 YEARS FRONT AROUND INJWELL−1
NUMBER OF POINTS= 46

X	Y	X	Y	X	Y	X	Y
0.586E+02	0.185E+03	0.608E+02	0.225E+03	0.588E+02	0.268E+03	0.520E+02	0.312E+03
0.397E+02	0.357E+03	0.216E+02	0.402E+03	−.266E+01	0.446E+03	−.334E+02	0.486E+03
−.703E+02	0.523E+03	−.113E+03	0.554E+03	−.161E+03	0.579E+03	−.212E+03	0.596E+03
−.266E+03	0.604E+03	−.321E+03	0.603E+03	−.376E+03	0.593E+03	−.428E+03	0.573E+03
−.475E+03	0.544E+03	−.517E+03	0.508E+03	−.552E+03	0.464E+03	−.579E+03	0.415E+03
−.596E+03	0.362E+03	−.604E+03	0.308E+03	−.603E+03	0.253E+03	−.593E+03	0.199E+03
−.574E+03	0.148E+03	−.547E+03	0.102E+03	−.514E+03	0.605E+02	−.476E+03	0.251E+02
−.435E+03	−.400E+01	−.391E+03	−.268E+02	−.346E+03	−.433E+02	−.301E+03	−.541E+02
−.257E+03	−.597E+02	−.215E+03	−.607E+02	−.175E+03	−.575E+02	−.138E+03	−.508E+02
−.104E+03	−.410E+02	−.735E+02	−.281E+02	−.462E+02	−.124E+02	−.221E+02	0.617E+01
−.124E+01	0.278E+02	0.166E+02	0.527E+02	0.316E+02	0.808E+02	0.437E+02	0.112E+03
0.528E+02	0.147E+03	0.586E+02	0.185E+03				

LINES TO FORM THE 4.0000 YEARS FRONT AROUND INJWELL−1
NUMBER OF POINTS= 46

X	Y	X	Y	X	Y	X	Y
0.212E+03	0.173E+02	0.213E+03	0.788E+02	0.212E+03	0.144E+03	0.207E+03	0.213E+03
0.195E+03	0.284E+03	0.176E+03	0.356E+03	0.148E+03	0.427E+03	0.111E+03	0.495E+03
0.624E+02	0.558E+03	0.421E+01	0.613E+03	−.631E+02	0.659E+03	−.138E+03	0.692E+03
−.218E+03	0.712E+03	−.301E+03	0.717E+03	−.383E+03	0.706E+03	−.462E+03	0.679E+03
−.535E+03	0.638E+03	−.598E+03	0.583E+03	−.649E+03	0.517E+03	−.687E+03	0.443E+03
−.710E+03	0.363E+03	−.717E+03	0.280E+03	−.708E+03	0.198E+03	−.685E+03	0.119E+03
−.648E+03	0.455E+02	−.600E+03	−.197E+02	−.543E+03	−.754E+02	−.478E+03	−.121E+03
−.409E+03	−.156E+03	−.338E+03	−.182E+03	−.266E+03	−.199E+03	−.195E+03	−.208E+03
−.127E+03	−.213E+03	−.630E+02	−.213E+03	−.268E+01	−.212E+03	0.526E+02	−.211E+03
0.102E+03	−.211E+03	0.144E+03	−.211E+03	0.177E+03	−.211E+03	0.198E+03	−.207E+03
0.209E+03	−.194E+03	0.211E+03	−.170E+03	0.211E+03	−.134E+03	0.210E+03	−.903E+02
0.211E+03	−.393E+02	0.212E+03	0.173E+02				

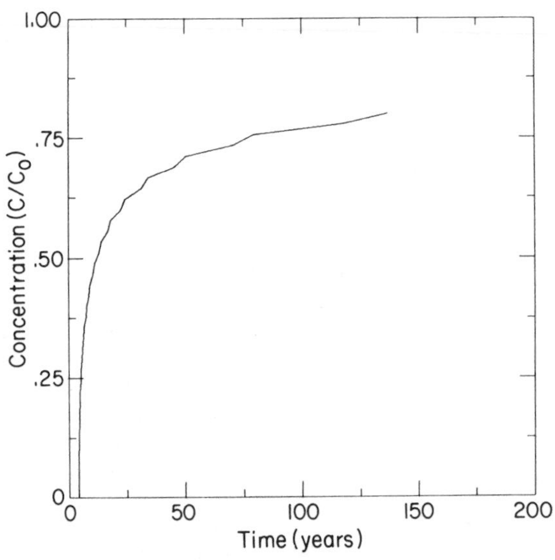

Fig. 17. Variation of dimensionless concentration C/C_o with time at the production well for example 1.

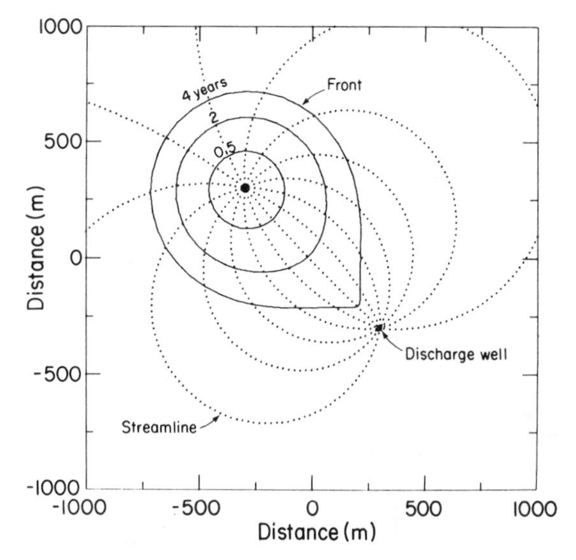

Fig. 18. Streamline pattern and front positions for example 1.

oriented in the direction perpendicular to the line connecting the two wells. Here the angle α in (65) is 45°. Given these conditions, the following items are required.

1. Show the pattern of streamlines leaving the injection well.
2. Show the streamline pattern for the whole system.
3. Draw the positions of the water-coincident contaminant front in the aquifer after 0.5, 2 and 4 years.
4. Determine how long it will take for a water-coincident particle to reach well B, the production well.
5. Present the time variation of contaminent concentration at the production well.

For this example, we will again use the RESSQ code. Input data will remain as given for example 1, with the following exceptions: columns 21-30 of Card 3 should now read 50. instead of 0. for regional flow velocity. The zero in columns 31-40 of Card 3 should be replaced by 45., which is the angle of regional flow with the positive x axis. An additional optional change would be to extend the boundaries under study from ± 1000. m to ± 2000. m.

Table 12 presents the arrival time of each streamline at the production well. Note that in this case only 19 of the 45 streamlines reach the production well during the 200-year period of study. The arrival time of the first streamline is 4.6 years, which is slightly longer than in example 1.

Figure 19 illustrates the flow pattern for those streamlines leaving the recharge well. The positions of the injected water front after 0.5, 2, and 4 years are also shown on Figure 19. Figure 20 presents the flow pattern for all streamlines, including those showing the uniform regional flow. To show the regional flow, injection wells with zero flow rate were placed in a row perpendicular to the direction of regional flow far from the injection/production well pair to act as regional flow streamline initiation points. One streamline leaves each zero-flow rate well, so their spacing along the row is determined by the requirement that streamline

TABLE 12. RESSQ output: arrival time of streamlines at the production well

STREAMLINES DEPARTING FROM INJECTION WELL INJWELL−1

NUMBER OF STREAMLINE	WELL REACHED	TIME OF ARRIVAL	ANGLE BETA IN DEGREES
1	+++NONE+++	200.0 YEARS	0.0
2	+++NONE+++	200.1 YEARS	8.0
3	+++NONE+++	200.1 YEARS	16.0
4	+++NONE+++	200.1 YEARS	24.0
5	+++NONE+++	200.0 YEARS	32.0
6	+++NONE+++	200.1 YEARS	40.0
7	+++NONE+++	200.1 YEARS	48.0
8	+++NONE+++	200.0 YEARS	56.0
9	+++NONE+++	200.2 YEARS	64.0
10	+++NONE+++	200.1 YEARS	72.0
11	+++NONE+++	200.2 YEARS	80.0
12	+++NONE+++	200.1 YEARS	88.0
13	+++NONE+++	200.1 YEARS	96.0
14	+++NONE+++	200.2 YEARS	104.0
15	+++NONE+++	200.0 YEARS	112.0
16	+++NONE+++	200.2 YEARS	120.0
17	+++NONE+++	200.1 YEARS	128.0
18	+++NONE+++	200.1 YEARS	136.0
19	+++NONE+++	200.2 YEARS	144.0
20	+++NONE+++	200.1 YEARS	152.0
21	+++NONE+++	200.1 YEARS	160.0
22	+++NONE+++	200.1 YEARS	168.0
23	+++NONE+++	200.1 YEARS	176.0
24	+++NONE+++	200.2 YEARS	184.0
25	+++NONE+++	200.1 YEARS	192.0
26	PRODWELL−1	21.5 YEARS	200.0
27	PRODWELL−1	11.6 YEARS	208.0
28	PRODWELL−1	8.7 YEARS	216.0
29	PRODWELL−1	7.2 YEARS	224.0
30	PRODWELL−1	6.2 YEARS	232.0
31	PRODWELL−1	5.6 YEARS	240.0
32	PRODWELL−1	5.1 YEARS	248.0
33	PRODWELL−1	4.8 YEARS	256.0
34	PRODWELL−1	4.7 YEARS	264.0
35	PRODWELL−1	4.6 YEARS	272.0
36	PRODWELL−1	4.7 YEARS	280.0
37	PRODWELL−1	4.8 YEARS	288.0
38	PRODWELL−1	5.0 YEARS	296.0
39	PRODWELL−1	5.4 YEARS	304.0
40	PRODWELL−1	6.1 YEARS	312.0
41	PRODWELL−1	7.0 YEARS	320.0
42	PRODWELL−1	8.5 YEARS	328.0
43	PRODWELL−1	11.4 YEARS	336.0
44	PRODWELL−1	21.9 YEARS	344.0
45	+++NONE+++	200.1 YEARS	352.0

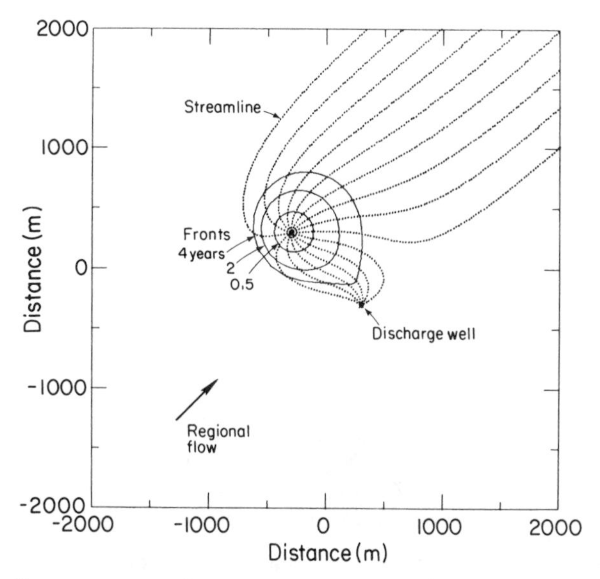

Fig. 19. Streamline pattern and front positions for example 2. Streamlines coming from regional flow are excluded.

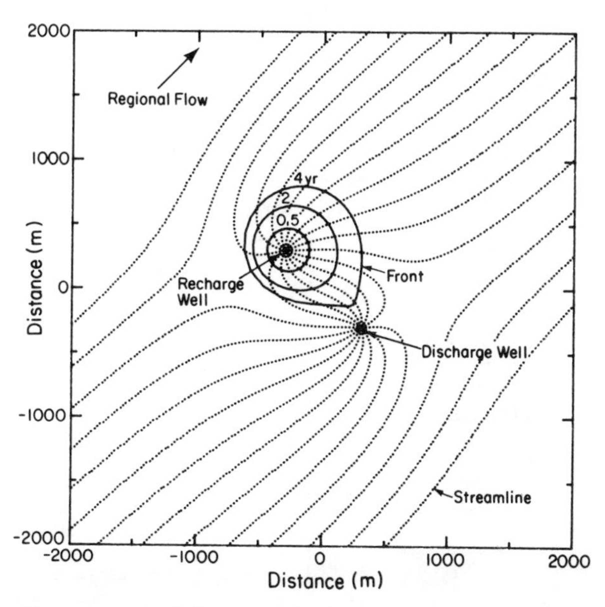

Fig. 20. Streamline pattern and front positions for example 2. Streamlines coming from regional flow are included.

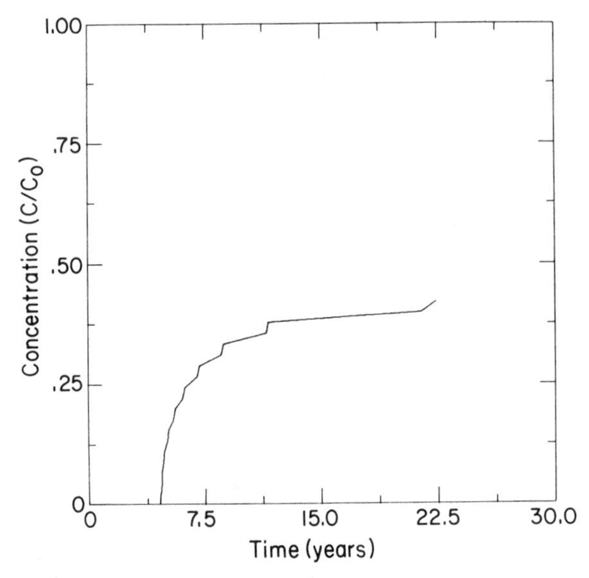

Fig. 21. Variation of dimensionless concentration C/C_o with time at the production well for example 2.

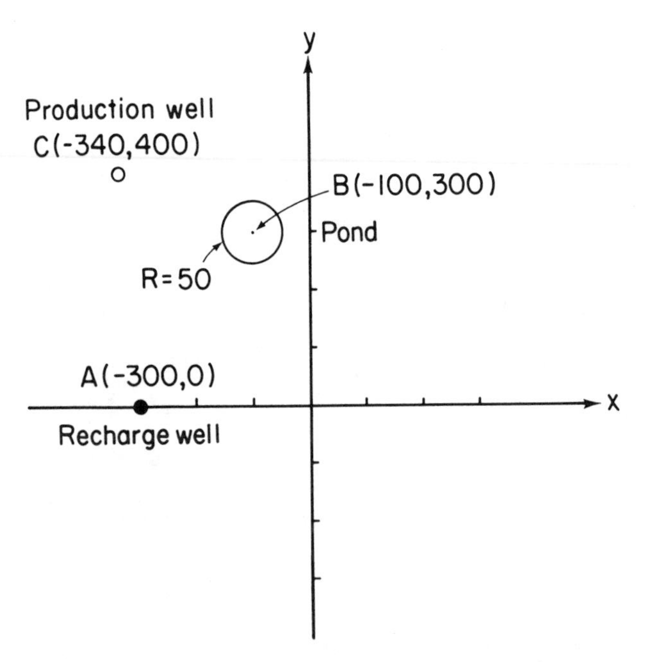

Fig. 22. A plan of location of different features in example 3.

density be proportional to fluid velocity (see Appendix H for a formula and coordinates of the zero-flow rate wells).

Figure 21 shows a plot of the variation in concentration of recharged water versus time at the production well. Notice that more than 50% of the stream tubes carrying the recharged water are washed away by the regional flow, leaving the final concentration at the production well below 50%.

3.5.3. *Example 3*

This example has been designed to demonstrate application of the semianalytical method to a more complex field problem that includes two different types of contaminant sources: a finite radius source representing a pond and a point source representing an injection well. For this example, let us consider a shallow aquifer 10 m thick with a regional flow making an angle of 315° with the positive x axis. The average pore water velocity of the regional flow is 62 m/yr. A liquid waste disposal pond with a diameter of 100 m fully penetrates the aquifer. The center of this pond has coordinates of (−100, 300) on the Cartesian system shown in Figure 22. The concentration of a particular chemical XX in the leachate leaving the pond is 5000 ppm. Liquid level in the pond is such that the volume rate of leachate leaving the pond is about 20 m^3/h. Liquid waste is also being recharged at a rate of 25 m^3/h into the aquifer through a well located at the point A shown in Figure 22. Concentration of XX in the recharged water of this well is 2000 ppm. A production well located upstream from both of these recharge sites (see Figure 22) is being continuously pumped at a rate of 30 m^3/h. We will assume the effective porosity of the aquifer to be 0.25. Given these conditions, we can perform the following steps.

1. Determine if the water from the production well contains any trace of the chemical XX. If so, determine the concentration of XX and its variation with time.
2. Show the streamline pattern of the system.
3. Plot the positions of the fronts after 0.33 and 1 year around both pond and recharge well.

Again the code RESSQ can be utilized. Data input to RESSQ may be prepared as was described in the previous examples. The finite radius pond may also be considered as a recharge well with a radius of 50 m.

Figure 23 illustrates the pattern of streamlines in the region encompassing the zone of interest. Fronts after 0.33 and 1 year around both pond and recharge well are also included. Note that several of the stream tubes leaving the pond and the recharge well arrive at the production well. Tables 13 and 14 present the arrival times of the streamlines reaching the production well from the pond and the recharge well, respectively. Figure 24 illustrates the time variation of the concentration of the chemical XX in the water of the production well. Table 15 shows the data from which Figure 24 was prepared.

3.5.4. *Example 4*

This example illustrates the use of a single pumping well for monitoring purposes, as was described in section 3.4. Consider a contaminant plume within a laterally infinite, homogeneous, isotropic, 10-m-thick aquifer with a porosity of 25%. To clean up the aquifer, the contaminated water is continuously pumped through a well fully penetrating the aquifer. To monitor the progress of the work, six observation wells are constructed around the pumping well as shown in Figure 25. While the pumping well operates at a constant rate of 45 m^3/h, the concentra-

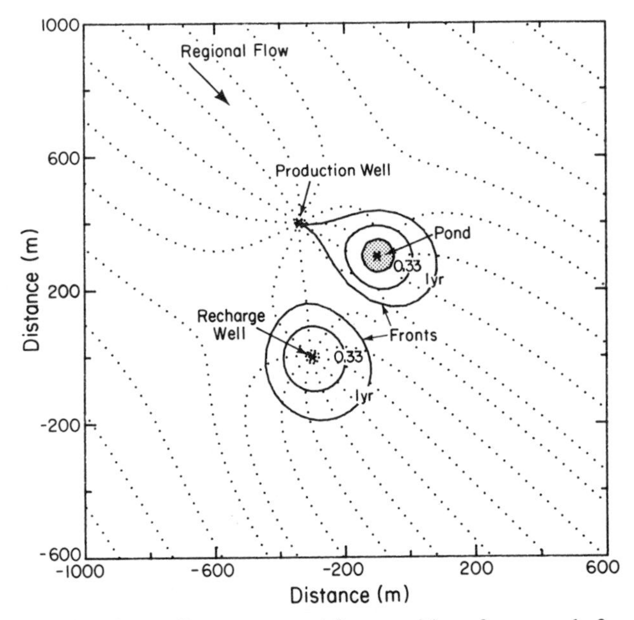

Fig. 23. Streamline pattern and front positions for example 3.

TABLE 13. RESSQ output: arrival time of streamlines at the producion well from the pond

STREAMLINES DEPARTING FROM INJECTION WELL POND−1

NUMBER OF STREAMLINE	WELL REACHED	TIME OF ARRIVAL	ANGLE BETA IN DEGREES
1	+++NONE+++	60.1 YEARS	0.0
2	+++NONE+++	60.1 YEARS	15.0
3	+++NONE+++	60.2 YEARS	30.0
4	+++NONE+++	60.1 YEARS	45.0
5	+++NONE+++	60.1 YEARS	60.0
6	PRODWELL−1	6.4 YEARS	75.0
7	PRODWELL−1	2.7 YEARS	90.0
8	PRODWELL−1	1.8 YEARS	105.0
9	PRODWELL−1	1.4 YEARS	120.0
10	PRODWELL−1	1.1 YEARS	135.0
11	PRODWELL−1	1.0 YEARS	150.0
12	PRODWELL−1	11.8 MONTHS	165.0
13	PRODWELL−1	1.0 YEARS	180.0
14	PRODWELL−1	1.1 YEARS	195.0
15	PRODWELL−1	1.3 YEARS	210.0
16	PRODWELL−1	1.7 YEARS	225.0
17	PRODWELL−1	2.8 YEARS	240.0
18	+++NONE+++	60.0 YEARS	255.0
19	+++NONE+++	60.1 YEARS	270.0
20	+++NONE+++	60.1 YEARS	285.0
21	+++NONE+++	60.1 YEARS	300.0
22	+++NONE+++	60.1 YEARS	315.0
23	+++NONE+++	60.0 YEARS	330.0
24	+++NONE+++	60.1 YEARS	345.0

TABLE 14. RESSQ output: arrival time of streamlines at the production well from the recharge well

STREAMLINES DEPARTING FROM INJECTION WELL INJWELL−1

NUMBER OF STREAMLINE	WELL REACHED	TIME OF ARRIVAL	ANGLE BETA IN DEGREES
1	+++NONE+++	60.1 YEARS	−6.0
2	+++NONE+++	60.1 YEARS	6.0
3	+++NONE+++	60.1 YEARS	18.0
4	+++NONE+++	60.1 YEARS	30.0
5	+++NONE+++	60.0 YEARS	42.0
6	+++NONE+++	60.0 YEARS	54.0
7	+++NONE+++	60.1 YEARS	66.0
8	PRODWELL−1	3.4 YEARS	78.0
9	PRODWELL−1	2.7 YEARS	90.0
10	PRODWELL−1	2.6 YEARS	102.0
11	PRODWELL−1	2.7 YEARS	114.0
12	PRODWELL−1	3.1 YEARS	126.0
13	PRODWELL−1	4.0 YEARS	138.0
14	PRODWELL−1	6.7 YEARS	150.0
15	+++NONE+++	60.1 YEARS	162.0
16	+++NONE+++	60.1 YEARS	174.0
17	+++NONE+++	60.0 YEARS	186.0
18	+++NONE+++	60.2 YEARS	198.0
19	+++NONE+++	60.1 YEARS	210.0
20	+++NONE+++	60.1 YEARS	222.0
21	+++NONE+++	60.1 YEARS	234.0
22	+++NONE+++	60.1 YEARS	246.0
23	+++NONE+++	60.0 YEARS	258.0
24	+++NONE+++	60.1 YEARS	270.0
25	+++NONE+++	60.0 YEARS	282.0
26	+++NONE+++	60.1 YEARS	294.0
27	+++NONE+++	60.1 YEARS	306.0
28	+++NONE+++	60.1 YEARS	318.0
29	+++NONE+++	60.0 YEARS	330.0
30	+++NONE+++	60.1 YEARS	342.0

TABLE 15. RESSQ output: time variation of concentration of chemical *XX* at the production well

EVOLUTION OF CONCENTRATION FOR PRODUCTION WELL PRODWELL−1

TIME IN MONTHS	CONCENTRATION IN PPM	(C−C0)/(CD−C0)
11.763	1.389E+02	0.0278
12.137	2.778E+02	0.0556
12.263	4.167E+02	0.0833
13.413	5.556E+02	0.1111
13.715	6.944E+02	0.1389
15.869	8.333E+02	0.1667
16.493	9.722E+02	0.1944
20.641	1.111E+03	0.2222
21.490	1.250E+03	0.2500
31.051	1.306E+03	0.2611
32.009	1.444E+03	0.2889
32.287	1.500E+03	0.3000
32.822	1.556E+03	0.3111

TABLE 15. (continued)

EVOLUTION OF CONCENTRATION FOR PRODUCTION WELL PRODWELL−1

TIME IN MONTHS	CONCENTRATION IN PPM	(C−C0)/(CD−C0)
33.967	1.694E+03	0.3389
36.793	1.750E+03	0.3500
40.403	1.806E+03	0.3611
47.410	1.861E+03	0.3722
77.347	2.000E+03	0.4000
80.690	2.056E+03	0.4111
361.085	2.056E+03	0.4111
365.320	2.056E+03	0.4111
370.551	2.056E+03	0.4111
384.422	2.056E+03	0.4111
400.969	2.056E+03	0.4111
430.556	2.056E+03	0.4111

TABLE 16. Time series of concentration data (ppm) used in example 4

	Well					
t hours	1	2	3	4	5	6
0	500	1000	1000	1000	--	1000
2.8	--	--	1000	--	--	1000
5.5	400	950	900	800	800	950
8	--	--	800	--	700	900
11	350	900	700	600	600	800
14	--	--	700	--	500	600
17	300	700	650	400	400	550
19	--	--	600	--	300	--
22	275	500	550	200	200	--
25	--	--	500	--	--	--
28	250	300	450	--	--	--

TABLE 17. Spatial concentration distributions calculated by program RT

$t = 0$								
x	y	C	x	y	C	x	y	C
−12.0	−1.0	500	4.7	−9.5	900	7.2	20.7	800
−13.2	−1.1	400	5.0	−10.0	800	7.4	21.0	700
−14.4	−1.2	350	5.4	−10.7	700	7.5	21.4	600
−15.5	−1.3	300	5.7	−11.3	700	7.6	21.7	500
−16.4	−1.4	275	6.0	−11.9	650	7.7	22.1	400
−17.4	−1.5	250	6.2	−12.3	600	7.8	22.3	300
−5.0	5.0	1000	6.4	−12.8	550	7.9	22.6	200
−6.4	6.4	950	6.7	−13.4	500	−8.0	−8.0	1000
−7.5	7.5	900	6.9	−13.9	450	−8.5	−8.5	1000
−8.6	8.6	700	15.0	5.0	1000	−8.9	−8.9	950

TABLE 17. (continued)

x	y	C	x	y	C	x	y	C
$t = 0$								
−9.4	9.4	500	15.9	5.3	800	−9.3	−9.3	900
−10.3	10.3	300	16.8	5.6	600	−9.8	−9.8	800
4.0	−8.0	1000	17.7	5.9	400	−10.2	−10.2	600
4.4	−8.8	1000	18.4	6.1	200	−10.6	−10.6	550
$t = 14$ hours								
−8.	0−.7	500	2.5	−5.0	900	6.6	18.9	800
−9.8	−.8	400	3.0	−6.0	800	6.7	19.2	700
−11.3	−.9	350	3.5	−7.1	700	6.9	19.6	600
−12.7	−1.1	300	4.0	−8.0	700	7.0	20.0	500
−13.8	−1.2	275	4.4	−8.8	650	7.1	20.4	400
−15.0	−1.3	250	4.7	−9.3	600	7.2	20.6	300
0.	0.	1000	5.0	−10.0	550	7.4	21.0	200
−.8	.8	950	5.4	−10.7	500	−4.9	−4.9	1000
−4.1	4.1	900	5.7	−11.3	450	−5.7	−5.7	1000
−5.8	5.8	700	12.4	4.1	1000	−6.3	−6.3	950
−6.9	6.9	500	13.5	4.5	800	−6.8	−6.8	900
−8.1	8.1	300	14.5	4.8	600	−7.4	−7.4	800
0.	0.	1000	15.5	5.2	400	−8.0	−8.0	600
1.8	−3.6	1000	16.3	5.4	200	−8.5	−8.5	550
$t = 28$ hours								
0.	0.	500	0.	0.	900	5.9	16.9	800
−4.0	−.3	400	0.	0.	800	6.0	17.3	700
−6.9	−.6	350	0.	0.	700	6.2	17.7	600
−9.0	−.8	300	0.	0.	700	6.3	18.1	500
−10.5	−.9	275	1.8	−3.7	650	6.5	18.5	400
−12.0	−1.0	250	2.4	−4.8	600	6.6	18.8	300
0.	0.	1000	3.0	−6.0	550	6.7	19.2	200
0.	0.	950	3.5	−7.1	500	0.	0.	1000
0.	0.	900	4.0	−8.0	450	0.	0.	1000
0.	0.	700	9.0	3.0	1000	0.	0.	950
−2.8	2.8	500	10.4	3.5	800	−2.6	−2.6	900
−5.0	5.0	300	11.7	3.9	600	−3.9	−3.9	800
0.	0.	1000	13.0	4.3	400	−4.9	−4.9	600
0.	0.	1000	13.9	4.6	200	−5.7	−5.7	550

C is in ppm, x and y are in meters.

TABLE 18. Input data for program RT for example 4

Card	Entry	Description
1	0.,0.	(x, y) coordinates of the pumping well (m)
2	10.,0.25,−45.,0.	Aquifer thickness (m), porosity, pumping rate (m^3/h), adsorption capacity of rock matrix
3	3	Number of concentration distribution "snapshots" to calculate
4	0.,14.,28.	Times of snapshots (hours)
5	3	Number of concentration contour levels to plot
6	100.,500.,900.	Concentration contour levels (ppm)
7	6	Number of observation wells
8	0	No concentration data from pumping well
9	−20.,20.,−15.,25.	x and y limits of the plotted area (m)
10	−12.,−1.	(x, y) coordinates of observation well 1 (m)
11	6	Number of (t, C) data pairs for observation well 1
12	0.,500.,5.5,400.,11., 350.,17.,300.,22., 275.,28.,250.	(t, C) data pairs for observation well 1 (hours, ppm)
13	−5.,5.	(x, y) coordinates of observation well 2 (m)
14	6	Number of (t, C) data pairs for observation well 2
15	0.,1000.,5.5,950.,11., 900.,17.,700.,22., 500.,28.,300.	(t, C) data pairs for observation well 2 (hours, ppm)
16	4.,−8.	(x, y) coordinates of observation well 3 (m)
17	11	Number of (t, C) data pairs for observation well 3
18	0.,1000.,2.8,1000.,5.5, 900.,8.,800.,11., 700.,14.,700.,17., 650.	(t, C) data pairs for observation well 3 (hours, ppm)
19	19.,600.,22.,550.,25., 500.,28.,450.	Remainder of (t, C) data pairs for observation well 3 (hours, ppm)
20	15.,5.	(x, y) coordinates of observation well 4 (m)
21	5	Number of (t, C) data pairs for observation well 4
22	0.,1000.,5.5,800.,11., 600.,17.,400.,22., 200.	(t, C) data pairs for observation well 4 (hours, ppm)
23	7.,20.	(x, y) coordiantes for observation well 5 (m)
24	7	Number of (t, C) data pairs for observation well 5
25	5.5,800.,8.,700.,11., 600.,14.,500.,17., 400.,19.,300., 22, 200.	(t, C) data pairs for observation well 5 (hours, ppm)
26	−8.,−8.	(x, y) coordinates of observation well 6 (m)
27	7	Number of (t, C) data pairs for observation well 6
28	0.,1000.,2.8,1000.,5.5, 950.,8.,900.,11., 800.,14.,600., 17., 550.	(t, C) data pairs for observation well 6 (hours, ppm)

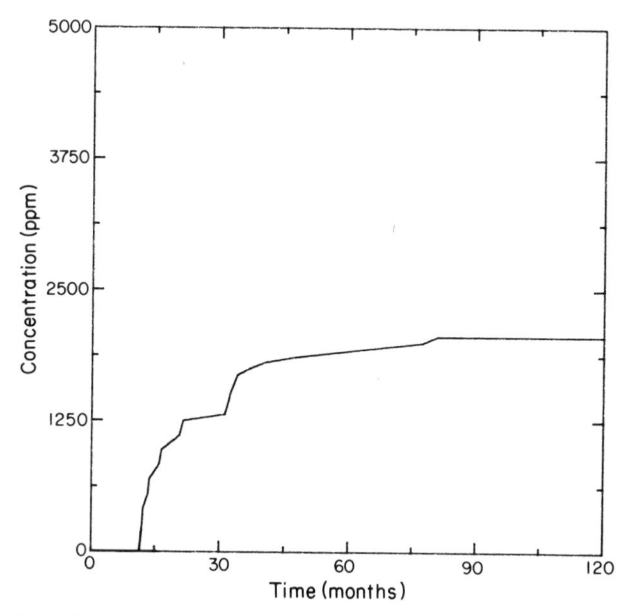

Fig. 24. Variation of contaminant concentration with time at the production well for example 3.

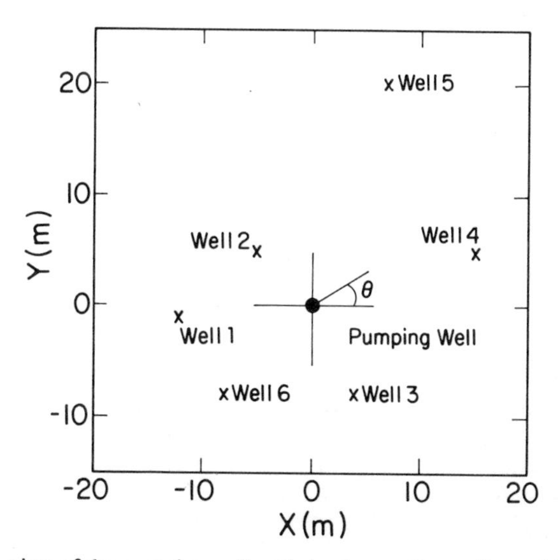

Fig. 25. Plan view of the pumping well and six observation wells used in example 4.

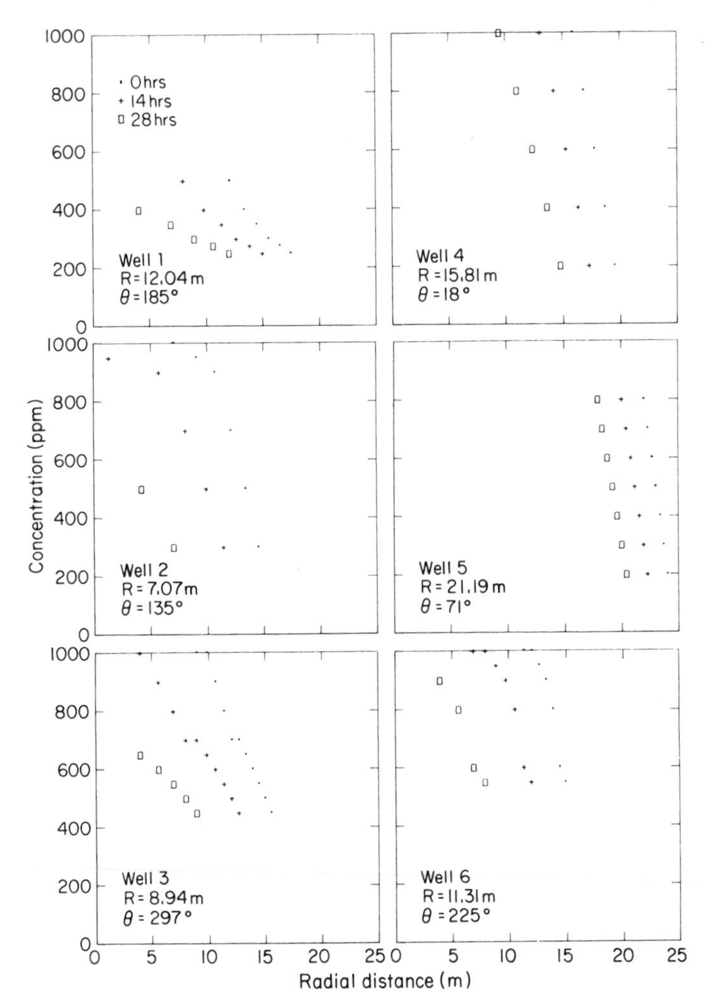

Fig. 26. Contaminant concentration as a function of radial distance from the pumping well for three times, calculated using computer program RT or (117). Angle θ is measured counterclockwise from the positive x axis.

tion of a contaminant *XX* is measured in each observation well at a series of times. The resulting time series of concentration data is given in Table 16.

Because the fluid flow in the aquifer is purely radial toward the pumping well, the following equation, which is given in section 3.4, may be used to convert the time-series of concentration measurements to a spatial concentration distribution:

$$r_2 = \left[r_1^2 - \frac{Q(t_2 - t_1)}{n\pi b} \right]^{1/2} \tag{117}$$

where

Q pumping rate, 45 m^3/h;
n porosity, 0.25;
b aquifer thickness, 10 m.

Note that dispersion and diffusion effects are not included in this method.

If t_2 is the desired time of the concentration distribution "snapshot", r_1 is the distance from the production well to a particular observation well, and t_1 is the time the concentration C_1 was measured in that observation well, then r_2 is the distance from the pumping well to the concentration C_1 (measured in the direction of that observation well) at time t_2. Each (t, C) data pair that makes up the time series of contaminant measurements for a given observation well thus yields an (r, C) data pair for the time t_2. The resulting concentration versus radial distance profile is plotted in Figure 26 for the observation wells for three values of t_2: 0, 14, and 28 hours after pumping begins.

By combining data from various observations wells (with different values of r_1 and different directions from the pumping well) a spatial distribution of contaminant concentration in the aquifer can be determined. The resulting (x, y, C) data, for three times, is given in Table 17. Figure 27 shows simple contour plots of the concentration distributions given in Table 17. A solid contour line indicates that the contour value was interpolated from the data, a dashed line indicates that the contour value was extrapolated from the data, and a dotted line indicates that that particular contaminant concentration reached the pumping well.

All of the results shown in this section were calculated using the computer program RT (Appendix I). The input required for RT is described in Table 18.

3.6. Advantages of Semianalytical Methods

The following are some of the advantages of semianalytical methods.

1. In the presence of multiple sources of contamination and discharge features such as pumping wells and effluent streams, where direct analytical solutions are not tractable, semianalytical methods can be used to estimate the order of magnitude of contamination extent for particular solute species.
2. For preliminary studies with limited budget and time or limited data available, semianalytical techniques are invaluable for estimating the travel time of a water-coincident or adsorption-retarded solute to a discharge well.
3. Application of these methods requires only simple computer input data and does not require the design of a mesh as with fully numerical methods.
4. An initial study using semianalytical methods can indicate whether or not a more sophisticated study based on a long period of observation and expensive data collection is required.

3.7. Limitations of Semianalytical Methods

The following are some of the limitations of semianalytical methods.

1. Semianalytical methods as discussed in this chapter do not consider mass transport by dispersion and diffusion, which in many cases may lead to the prediction of travel times which are longer than actual values and may underestimate the true impact of a contaminant source.
2. Since development of the technique is based on a two-dimensional plane

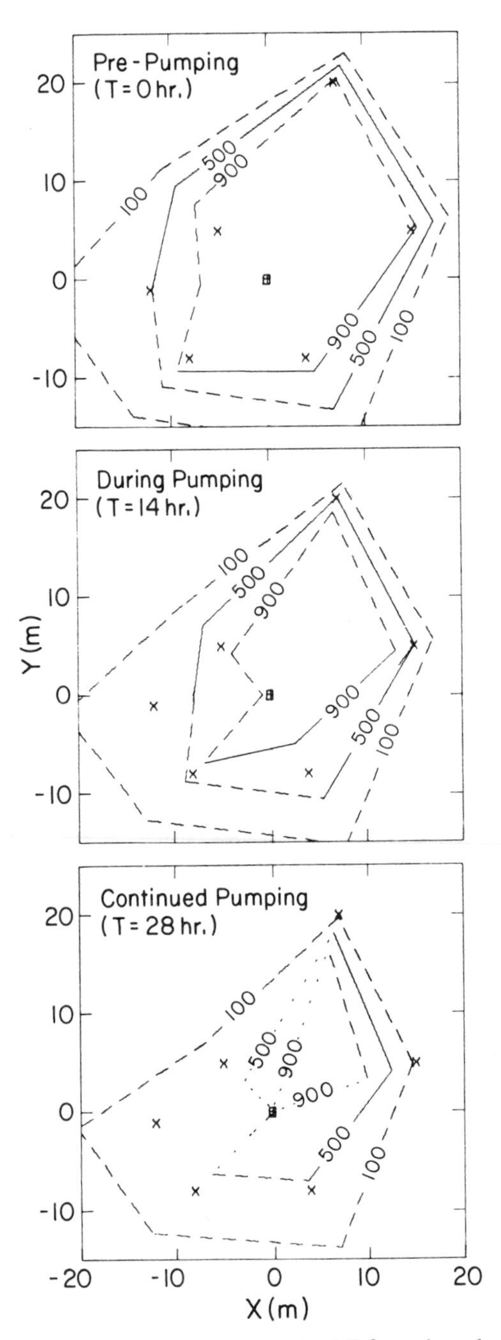

Fig. 27. Concentration contours calculated by RT from the values in Table 17.

theory, field problems that are actually three-dimensional in nature must be simplified before semianalytical methods can be applied.

3. Semianalytical methods cannot handle media with heterogeneous or anisotropic permeabilities.

4. The methods described in this chapter hold only for steady state problems, although in some cases they could be extended to handle transient problems.

4 Numerical Methods

Numerical methods are generally required to solve complex equations describing coupled or uncoupled processes in heterogeneous and anisotropic formations under various initial and boundary conditions. In most numerical models the governing equations are approximated by algebraic equations relating unknown variables at discrete nodal points and at different times. The governing equations for fluid flow and solute transport have second-order diffusive terms $\nabla\cdot(\mathsf{K}\,\nabla h)$ and $\nabla\cdot(\mathsf{D}\,\nabla C)$, a first-order convective term $\mathbf{v}\cdot\nabla C$; and transient terms $\partial h/\partial t$ and $\partial C/\partial t$. The accuracy and efficiency of a model depend (1) on the numerical approximations for evaluating the spatial gradient ∇ and the time derivative $\partial/\partial t$ and (2) on the solution scheme of the algebraic equations.

This section discusses general features of numerical methods and different approaches used to determine fluid flow field and solute concentration distribution. The finite difference, finite element, and flow path network methods, explicit and implicit time-stepping schemes, and iterative and direct equation solvers are briefly described. The characteristics of a number of different numerical codes are then summarized in Tables 19, 20, and 21. The code listing in this study is based on a literature review and extends previous surveys sponsored by EPA [*Bachmat et al.*, 1978], Department of Energy (DOE) [*Science Applications, Inc.*, 1981; *Wang et al.*, 1983], Nuclear Regulatory Commission (NRC) [*Thomas et al.*, 1982], and *International Ground Water Modeling Center* [1983].

4.1. Spatial Approximations

To calculate fluid pressure and solute concentration, the finite difference, integrated finite difference, and finite element methods are frequently used to approximate first- and second-order spatial derivative terms. The main distinctions among the different methods are in the numerical approximation of the gradient operator ∇, the evaluation of variable-dependent coefficients, and the spatial discretization of the region. For the modeling of complex geological formations, it is important to be able to handle the large number of equations that result from an irregular discretization of multidimensional space.

4.1.1. *Finite Difference Method*

In most finite difference models, the distribution of nodes is regular, creating a grid with either uniform or nonuniform spacing along orthogonal coordinate systems (Cartesian: x, y, z; cylindrical: r, θ, z; etc.). Surrounding each nodal point there is a region bounded by interfaces normal to the coordinate axes; this region is called a nodal block, cell, or element. Between two nodes indexed by i and i+1 in the x direction, the interface i+1/2 can intersect the x axis either midway between i and i+1, or at an off-center location.

For the evaluation of a spatial gradient the partial differential of a variable is expressed in terms of the difference between two neighboring nodal values. For example, the x component of the concentration gradient $\partial C/\partial x$ at the interface $i \pm 1/2$ is approximated by

$$\frac{C_{i\pm1} - C_i}{x_{i\pm1} - x_i} \tag{118}$$

With the finite difference approximation, the nodal value of C_i is algebraically related to its two neighboring values for a one-dimensional problem, or six neighbors for a three-dimensional problem. The components of the second-order terms' coefficients K and D, at the interfaces $i \pm 1/2$, can be evaluated as the arithmetic mean,

$$D_{i\pm1/2} = \frac{D_{i\pm1} + D_i}{2} \tag{119}$$

or as the harmonic mean,

$$\left(\frac{1}{D}\right)_{i\pm1/2} = \frac{(\Delta x/D)_{i\pm1} + (\Delta x/D)_i}{\Delta x_{i\pm1} + \Delta x_i} \tag{120}$$

where

$$\Delta x_i = |x_{i\pm1/2} - x_i| \qquad \Delta x_{i\pm1} = |x_{i\pm1/2} - x_{i\pm1}|$$

These approximations for interface values can be generalized for an irregular distribution of nodes. For example, the factor 1/2 in the arithmetic mean can be replaced by other fractional weighting factors; the Δx_i in the harmonic mean can be replaced by the normal distance from the nodal point to the interface.

For the first-order convective term $\mathbf{v} \cdot \nabla C$ in the solute transport equation, $\partial C/\partial x$ may be approximated by the central difference in space, or central weighting,

$$\frac{C_{i+1/2} - C_{i-1/2}}{x_{i+1/2} - x_{i-1/2}} \tag{121}$$

with $C_{i\pm1/2} = 0.5\,(C_{i\pm1} + C_i)$. With this central weighting scheme, the space truncation approximation of the convective term is correct to second order. However, there is a tendency for solutions with central weighting to oscillate artificially at high flow velocities. The convective flux associated with the flow velocity $\mathbf{v}$ carries solute downstream; a nodal point between an upstream and a downstream node will have a concentration closer to the upstream value. Since the central weighting scheme does not take into account this convective effect, other schemes have been devised. In the upstream weighting scheme, the interface concentration is set equal to the upstream value: that is, $C_{i+1/2} = C_{i+1}$ if fluid flows from i+1 to i. In other words, a backward difference, $C_{i+1} - C_i$, is used for the convective term at node i. The upstream weighting eliminates the oscillation but introduces a space discretization error, called numerical dispersion, which produces the same effect as physical dispersion. The error of numerical dispersion in the upstream weighting scheme and the error of numerical oscillation in the central weighting scheme may be minimized by using the partial upstream weighting $C_{i+1/2} = aC_{i+1} + (1 - a)C_i$ with $0.5 \leqslant a \leqslant 1$, or discontinuous weighting, with central weighting at low flow velocity and upstream weighting at high flow velocity. Other weighting schemes have also been devised.

4.1.2. *Integrated Finite Difference Method*

The integrated finite difference method (IFDM) is a more flexible version of the finite difference method. In the integrated finite difference method, the distribution of nodal points may form an irregular mesh, and the modal blocks may be arbitrarily shaped polyhedrons. The numerical equations are formulated from the integral form of the governing equations, as opposed to the simpler finite difference methods that employ the differential form of the governing equations. The IFDM formulation emphasizes the direct representation of the conservation laws in relating the rates of change of fluid mass and solute mass in each nodal block to the fluxes over the interfaces bounding that block [*Edwards*, 1972; *Narasimhan and Witherspoon*, 1976]. To evaluate the rates of change and the fluxes over the boundary surfaces of a block, the volume, surface areas, and normal distances from the node to the faces of the polyhedron are required input data to be specified by the modeler. These additional input requirements allow great flexibility in the mesh design for complex geometry systems but increase data requirements and programming load considerably. For a regular mesh in orthogonal coordinates, the IFDM yields results essentially equivalent to those generated by the finite difference method. Both methods use simple finite differencing, or linear interpolation between neighboring nodal values, in the evaluation of spatial gradients normal to the interfaces.

4.1.3. *Finite Element Method*

The finite element method, like the IFDM, has the flexibility of specifying the distribution of nodes and using an irregular mesh to divide the region into elements. In the finite element method an element is the region bounded by curves connecting the nodes. Different element shapes can be defined. For example, a two-dimensional linear triangular element is a triangle with three nodal vertices, a quadrilateral element has four corner nodes, and a three-dimensional orthorhombic element has eight corner nodes [*Zienkiewicz*, 1977; *Pinder and Gray*, 1977]. Within a model, different types of elements can be used. In two-dimensional problems, fractures can be treated with one-dimensional line elements, while the porous medium blocks can be represented with triangular or quadrilateral elements.

The value of a variable within an element is interpolated in terms of the values of the variable at the corner nodes. Simple polynomials (linear, quadratic, or cubic) are frequently used as linearly independent basis functions for the interpolation. If both state and space variables in a problem are interpolated with the same function, the element is referred to as an isoparametric element. For linear interpolation the values at the corner nodes are sufficient to define the basis functions for the interpolation. For quadratic or cubic interpolations the basis functions are specified with either the values at additional side nodes or the values of the partial derivatives of the variable at the corner nodes. For example, the three-dimensional Hermite interpolation functions are a set of four cubic polynomials defined by the value and its three partial derivatives at each corner node.

The finite element numerical equations are usually formulated with either the weighted residual Galerkin scheme or the variational approach. In the Galerkin finite element scheme, a trial solution made up of an expansion of basis functions is substituted into the differential equations. The space differential operators operate on the basis functions. The residual of the trial solution is integrated over the element, weighted by the same basis functions. The integration is usually carried out using two- or three-point Gaussian integration for each dimension. If the trial solution were to be expanded in terms of a complete set of an infinite number of linearly independent basis functions, the trial solution would be exact and the residuals would vanish. In the Galerkin method the number of basis functions is finite and the residuals are forced to be zero by requiring orthogonality of the residuals to the set of basis functions used in the trial solution. For the convective terms the problems of numerical oscillation and numerical dispersion also exist in the finite-element method. Upstream basis functions can be used to minimize these effects.

An equivalent expression of the governing partial differential equations can be given in terms of variation of functionals. A functional is a function of functions such as an integral over space with the integrand composed of basis functions. Upon minimization of an appropriate functional the corresponding differential equations emerge. The variational approach for fluid flow or solute transport is based on the same minimum energy principle or Lagrangian formulation as that used to study the equilibrium states in mechanics or stress analysis. In the variational approach to the finite element method, the trial solutions, expansions in basis functions, are substituted into the functional integrals. The differential operators in the functional integrals operate on the basis functions in a manner similar to that in the weighted residual procedure of the Galerkin formulation.

With the use of the Gaussian algorithm for element integration, the coefficients K and D in the second-order terms and $\mathbf{v}$ in the convection term are evaluated at the Gaussian points within an element. This is different from the finite difference method with the coefficients calculated at the interfaces between blocks. The finite element method, with the use of the basis function interpolation over more than two points, can evaluate gradients in both normal and tangential directions and handle tensorial quantities more easily.

4.1.4. *Flow Path Network Method*

Instead of solving the governing differential equation for solute transport, the solute concentration can also be determined by calculating the motion of a large number of discrete solute particles. At each time step the new position of a solute particle is determined by the fluid particle velocity, the retardation factor and the dispersivity. Each particle is also assigned a weight which can be changed at each time step to account for radioactive decay, creation of new daughter nuclides, or chemical reactions. With the position and weight of each particle varying over time, the concentration of each species is calculated for a set of "cells" by summing the weight of the particles of that species in each cell and dividing by the volume of water in the cell.

Different schemes can be used to account for solute dispersion in determining the position of each particle. One approach is to use random numbers uniformly distributed between -0.5 and $+0.5$ to determine the forward or backward net dispersive displacement in each time step [*Ahlstrom et al.*, 1977; *Schwartz and*

Crowe, 1980]. With the use of a large number of discrete point particles, the random number approach may adequately represent solute dispersion. Another approach is to consider a given distribution (e.g., Gaussian) for solute particle velocity. Dispersion is treated by dividing the solute in each cell into packets with different velocities which are chosen so as to divide the velocity distribution into intervals of equal area [*Campbell et al.*, 1981].

By directly solving for the position changes of solute particles or packets instead of approximating the differential equation for concentrations, the flow path method does not have the numerical dispersion problems encountered in the finite difference and finite element methods. However, the flow path method may be computationally less efficient and is generally limited to simple flow patterns so that the flow path can be easily traced in a simple network. Generally a flow path network code is a complement to a finite difference or finite element code.

4.2. **Temporal Approximations**

First-order finite difference in time is frequently used to approximate the time derivative $\partial/\partial t$ for the transient rates of change in the governing equations. According to the conservation law the rates of change of fluid and solute mass are equated to the fluxes and the source/sink terms. Thus the temporal approximations should be chosen in accordance with the spatial methods adopted in a model.

In the integrated finite difference and finite difference methods, the value of the variable at a given node represents the average value within the block enclosed by the interfaces on which the fluxes are evaluated. The balance between the rate of accumulation within the block and the net flux across the interfaces is explicitly preserved in the numerical equations. In the finite element method, each element is shared by the nodes on the boundary and each node is surrounded by several elements. Although the transient term can be handled easily in the finite element formulation, the mathematical relationship between the rate of accumulation associated with a block and the fluxes evaluated at the Gaussian points in the surrounding elements is an indirect representation of the conservation law.

4.2.1. *Implicit Equations*

With either the finite difference or the finite element method the analysis of the transient equation results in a system of equations of the matrix form

$$[A]\left\{\frac{df}{dt}\right\} + [B]\{f\} + \{R\} = 0 \tag{122}$$

where the column $\{f\}$ contains the nodal values of pressure and solute concentrations. The coefficient matrix $[A]$ contains the coefficients of the fluid and solute storage capacity associated with the time derivative $\{df/dt\}$, $[B]$ contains the spatial approximations (finite difference or finite element) of the fluxes, and $\{R\}$ contains the known information such as source/sink or boundary conditions.

The first-order temporal finite difference from time t to $t + \Delta t$ is

$$\left\{\frac{df}{dt}\right\} \approx \frac{\{f\}_{t+\Delta t} - \{f\}_t}{\Delta t} \tag{123}$$

To solve for the unknown $\{f\}_{t+\Delta t}$ from the known solution $\{f\}_t$, the other terms in the governing equations can be interpolated between $t + \Delta t$ and t. With linear interpolation, the matrix equation becomes

$$\frac{[A]}{\Delta t}(\{f\}_{t+\Delta t} - \{f\}_t) + [B]\left(\lambda\{f\}_{t+\Delta t} + (1-\lambda)\{f\}_t\right)$$

$$+ \lambda\{R\}_{t+\Delta t} + (1-\lambda)\{R\}_t = 0 \tag{124}$$

For the forward differencing explicit scheme the interpolation factor λ equals zero, and $\{f\}_{t+\Delta t}$ can be easily determined by multiplying the matrix equation by $\Delta t[A]^{-1}$. The explicit scheme generally requires a minimum of computational effort. However, it is only conditionally stable. Usually, implicit schemes with interpolation factor $0.5 \leqslant \lambda \leqslant 1$ are stable. The central differencing Crank-Nicholson scheme ($\lambda = 0.5$) is accurate in Δt to second order. The backward differencing implicit scheme ($\lambda = 1.0$) is usually unconditionally stable and is correct in time to first order.

4.2.2. *Coupling Solution Schemes*

The coupled equations of pressure and concentration can be solved either sequentially or simultaneously. The sequential method solves the equations separately and treats the variables as unknowns only when their respective equations are being solved. The fluid flow equation for pressure is solved first. Then the transport equation for concentration is solved using the velocity field calculated from the Darcy equation based on the new values of pressure. The coupled equations can also be solved simultaneously, which involves larger matrices and therefore greater computer time and storage requirements.

4.3. **Matrix Solvers**

After temporal finite differencing and spatial discretization have been done, the partial differential equations are transformed into a system of simultaneous linear algebraic equations or a matrix equation of the form

$$[M]\{f\} = \{F\} \tag{125}$$

The size of the matrix depends on the number of nodes, the number of variables, and the solution schemes. The solution for the unknown nodal values in $\{f\}$ at time $t + \Delta t$ can be obtained through the use of either direct elimination methods or iterative methods.

4.3.1. *Direct Elimination Methods*

Many of the direct elimination methods are variations of the Gaussian elimination procedure. In this procedure, one unknown is eliminated from one equation at a time (forward elimination). The procedure works in a systematic way so that a general matrix equation is reduced to a triangular system. In this triangular system, the last equation has one unknown, the second-to-last equation has two unknowns, etc. The triangular system can be solved step by step, the last (nth) unknown being determined by the last equation and then the ($n-1$)th unknown being determined by the ($n-1$)th equation upon substitution of the nth solution. This backward substitution proceeds until all the unknowns are determined.

L-U (lower-upper) decomposition is one method of Gaussian elimination. With the matrix $[M]$ decomposed into a lower and an upper triangular matrix, $[M] = [L][U]$, the matrix equation $[M]\{f\} = \{F\}$ is equivalent to two triangular systems:

$$[L]\{g\} = \{F\} \qquad [U]\{f\} = \{g\} \tag{126}$$

If $[L]$ and $[U]$ are known, the matrix equation can be solved by backward substitution. With a given $n \times n$ matrix $[M]$, the matrices $[L]$ and $[U]$ are not unique. There are n^2 elements in $[M]$ and $0.5n(n+1)$ unknown elements in each of $[L]$ and $[U]$. Therefore there are $n(n+1) - n^2 = n$ elements which can be set to any value. In the Doolittle method the diagonal elements of $[L]$ are set to unity. With the n diagonal elements fixed, the other elements in $[L]$ and $[U]$ can be determined algebraically. Alternatively, in the Crout method the diagonal elements of $[U]$ are set to unity instead. The Crout and the Doolittle method are two popular direct elimination solution schemes [*Dahlquist and Bjorck*, 1974].

The efficiency of a direct matrix solver depends strongly on the structure of a matrix. For the tridiagonal banded matrix frequently encountered in the application of numerical methods, the number of algebraic operations is approximately $5n$, which is much smaller than the $n^3/3$ required for Gaussian elimination of a general $n \times n$ matrix [*Dahlquist and Bjorck*, 1974]. In the alternating direction implicit (ADI) finite difference method for a regular grid in two- or three-dimensional space, the partial differentials along different directions are solved and updated sequentially for fractional (1/2 or 1/3) time steps. Each nodal unknown for a quasi-one-dimensional fractional step is connected to only two neighboring unknowns, so that the matrix has nonzero elements only on the diagonal and two nearest off-diagonals. The matrix equation can then be easily solved by the tridiagonal or Thomas' algorithm.

In most two- or three-dimensional methods the matrix is sparse, with the number of nonzero off-diagonal matrix elements in each row or column depending on the number of neighbors of the corresponding node. The positions of the off-diagonal matrix elements relative to the diagonal elements depend on the ordering of the nodes. Different ordering or numbering schemes can be made either to optimize the banding of matrix elements or to express parts of the matrix in diagonal or triangular form and minimize the computational effort. Another procedure required to ensure numerical stability and to minimize round off errors is the pivoting operation, which interchanges one row with another row or one column with another column in the matrix. It is necessary to perform the pivoting operation if zero or nearly zero diagonal elements are found during the Gaussian elimination procedure.

4.3.2. *Iterative Methods*

If the matrix is sparse and large, iterative methods may offer certain advantages over direct methods. An iterative method starts from a first approximation which is successively improved until a sufficiently accurate solution is obtained. Some examples of iterative methods are briefly described here to demonstrate the procedure involved [*Dahlquist and Bjorck*, 1974].

If one splits a matrix $[M]$ into diagonal $[D]$, lower $[L]$, and upper $[U]$ triangular systems, $[M] = [D] + [L] + [U]$. The matrix equation $[M]\{f\} = \{F\}$ can be rewritten as

$$([D] + [L])\{f\} = -[U]\{f\} + \{F\} \tag{127}$$

An approximate solution from the kth iteration step to the $(k+1)$th step is

$$\{f\}^{k+1} = ([D] + [L])^{-1}(-[U]\{f\}^k + \{F\}) \tag{128}$$

This is the matrix form of Gauss-Seidel's method. The inverse of the lower triangular matrix $([D] + [L])$ can be handled by forward substitution.

If the residual from the kth to the $(k+1)$th iteration is denoted by $\{r\}^k = \{f\}^{k+1} - \{f\}^{2k}$, the generalized iterative method $\{f\}^{k+1} = \{f\}^k + w\{r\}^k$ is the successive overrelaxation (SOR) method. In matrix form the new solution is

$$\{f\}^{k+1} = ([D] + w[L])^{-1} \left[\Big((1 - w)[D] - w[U] \Big) \{f\}^k + w\{F\} \right] \tag{129}$$

The relaxation factor w should be chosen so that the rate of convergence is optimized. Eigenvalue analyses are frequently used for determining the best relaxation factor. For real, symmetric, and positive definite matrices, $0 < w < 2$. Other relaxation or acceleration schemes can be constructed in a manner similar to that used for the successive overrelaxation method.

The advantages of direct methods and iterative methods can be combined in the block iterative methods. In the block iterative methods the coefficient matrix is partitioned into blocks and all elements of a block are operated on during one iterative step. Within each block a direct solution scheme is used. The ADI procedure is an example of the block iterative method. Each block is tridiagonal and can be easily solved. In general, the block iterative method is superior to the corresponding point iterative method.

4.4. Computer Codes

With the formulation of governing equations and numerical methods, a specific computer code can be developed by constructing an algorithm, eliminating coding errors, running sample problems, and producing a user's guide. Many codes are initially developed to model a specific class of problems. When a code is used for other problems, modifications can be added to generalize the code capabilities. The versatility and efficiency of a code can also be improved by adopting better solution schemes and numerical methods. In many cases a code becomes a well-established and powerful tool as the result of efforts by both competent developers and experienced users.

With the growing concern over contaminant transport in the environment, in the past few years many codes for fluid flow and solute transport have been developed. Organizations such as EPA, NRC, and DOE, among others, have sponsored several surveys to review the capabilities of various codes. The focus of one EPA study [*Bachmat et al.*, 1978] is on water resource management. As a result of the EPA study, an International Ground Water Modeling Center has been established at Holcomb Research Institute (Butler University, Indianapolis) to continuously gather information and produce training programs for groundwater modeling. In addition to the concern over water quality, the need to predict radionuclide transport from underground repository construction and operation also contributes to the development of transport modeling. DOE and NRC also sponsor surveys on fluid flow and solute transport [e.g., *Science Applications, Inc.*, 1981; *Thomas et al.*, 1982].

Thus far we have reviewed the methodologies commonly used in numerical codes; Tables 19, 20, and 21 summarize the main characteristics of a number of finite difference, finite element, and flow path network codes, respectively. In each table, the codes are ordered according to the first author's name. Simulation examples, numerical characteristics, fluid flow processes, and solute transport processes are described in separate columns. Code documentation, when available, has been cited in the reference list according to first author's name and reference date. Often several versions of a code are developed to solve different types of problems; to avoid repetition, only one variation of any code is included in the tables.

In order to select an appropriate code to use in simulations, the user must first determine the processes to be modeled, the availability of data on fluid, solute, and formation properties, the complexity of the modeled regions, and the initial and boundary conditions. Use of analytical or semianalytical analysis in conjunction with complex numerical codes is extremely useful; it can lend insight into physical processes and help interpret numerical results. With modeling experience and the recent improvement in numerical codes, the user can model subsurface fluid flow and solute transport to aid in management of waste disposal, containment of existing contamination, and design of remedial actions for subsurface environmental protection.

4.5. Example of the Use of a Sophisticated Numerical Model

In this section we describe the use of the numerical model PT, which is part of the family of codes including TRUST (Table 19), developed at Lawrence Berkeley Laboratory [*Bodvarsson*, 1982]. PT uses the integrated finite difference method to calculate heat and mass transfer in a water-saturated porous or fractured medium. It can be used for one-, two-, or three-dimensional complex geometry problems involving heterogenous materials. Fluid density is temperature and pressure dependent and fluid viscosity is temperature dependent. The vertical deformation of the rock matrix may be calculated using the one-dimensional consolidation theory of *Terzaghi* [1925]. The code can handle temperature-dependent thermal conductivities and media with anisotropic permeability. The following physical effects may be included: (1) heat convection and conduction in the aquifer/aquitard system, (2) regional groundwater flow, (3) multiple heat and/or mass sources and sinks, (4) hydrologic or thermal barriers, (5) constant pressure or temperature boundaries, and (6) gravitational effects. PT was developed from the code CCC [*Lippmann et al.*, 1977], which has been used for many years for a variety of energy storage, geothermal, and waste isolation problems. PT employs a direct elimination method with L-U decomposition which is much more efficient than the iterative method used by CCC to solve the coupled mass and energy equations. Both PT and CCC have been validated against various analytical solutions and the results of several field experiments [*Bodvarsson*, 1982].

TABLE 19. Finite difference solute transport codes

Authors (Institute)	*Code Name* (Reference Date) Simulation Examples	Numerical Characteristics	Fluid Flow Processes	Solute Transport Processes
R. T. Dillon R. B. Lantz S. B. Pahwa (Sandia Natl. Lab.)	*SWIFT* (1978) Radionuclide transport	Three-dimensional Two line SOR iterative or direct ordered Gaussian elimination	Transient Well-bore Nonisothermal Convection	Advection Dispersion Diffusion Adsorption
G. R. Dutt M. J. Shaffer W. J. Moore (Bureau of Reclamation)	*Salt Transport in Irrigated Soils* (1972) Predict quality of irrigation return flow	One-dimensional vertical coupled with chemistry model	Transient Unsaturated Homogeneous, isotropic media	Ion exchange
Intera Environmental Consultants, Inc.	*HCTM* (1975) Tritium transport Mine dewatering needs	Three-dimensional Method of characteristics for advection	Saturated/ unsaturated	Advection Dispersion Diffusion Sorption; Decay
T. R. Knowles (Texas Dept. of Water Research)	*GWSIM-II* (1981)	Two-dimensional areal	Transient Confined/uncon-fined aquifer	Advection Dispersion Diffusion
L. E. Konikow J. D. Bredehoeft J. V. Tracy (U. S. Geological Survey)	*KONBRED*[a] *USGS-2D-Transport/MOC*[a] (1978; *Tracy*, 1982) Chloride movement, Rocky Mountain Arsenal, CO and stream-aquifer system Radionuclide transport INEL, ID	Two-dimensional areal ADI for flow equation Method of characteristics for solute transport equation	Transient Confined, semi-confined, water table aquifer Diffuse leakage Injection/with-drawal	Advection Dispersion Adsorption Decay

TABLE 19. (continued)

Authors (Institute)	*Code Name* (Reference Date) Simulation Examples	Numerical Characteristics	Fluid Flow Processes	Solute Transport Processes
J. A. Korver (Lawrence Livermore Lab.	*OGRE* (1970) Radionuclide transport around underground openings	Two-dimensional axisymmetric Three-dimensional[b] ADI	Transient Saturated/ unsaturated	Advection Dispersion[b] Adsorption[b]
E. Ledoux (Ecole des Mines de Paris)	*NEWSAM* (1976) Salt transport	Two-dimensional areal Heirarchical grid	Transient	Advection Adsorption
T. N. Narasimhan A. E. Reisenauer K. T. Key R. W. Nelson (Lawrence Berkeley Lab., Pacific Northwest Lab.)	*TRUST+ FLUX/MILTVL* (*Reisenauer*, 1982) Soil columns Field consolidation	IFDM; Three-dimensional Complex geometry Pathline solutions	Saturated/ unsaturated Deformable porous medium	Advection
L. Rickertsen (Science Appl. Inc.)	*MIGRAIN* (*Claiborne et al.*, 1980) Migration of brine inclusions in salt	Two-dimensional, Three-dimensional Upstream weighting	Heat-induced migration	
J. B. Robertson (U.S. Geological Survey)	*Robertson 1* (1974) Radioisotope migration, National Reactor Testing Station	Two-dimensional areal ADI Method of characteristics for transport	Transient Bounded aquifer	Advection Dispersion Sorption; Decay Ion exchange

TABLE 19. (continued)

Authors (Institute)	*Code Name* (Reference Date) Simulation Examples	Numerical Characteristics	Fluid Flow Processes	Solute Transport Processes
A. Runchal D. Treger G. Segal (Dames & Moore)	*GWTHERM* (1979) Repository studies	Two-dimensional vertical IFDM; ADI Particle tracking option	Transient Nonisothermal Convection	Advection Diffusion Adsorption Decay

[a]Alternate names for the same code
[b]Features added after 1970

TABLE 20. Finite element solute transport codes

Authors (Institute)	*Code Name* (Reference Date) Simulation Examples	Numerical Characteristics	Fluid Flow Processes	Solute Transport Processes
R. G. Baca R. C. Arnett I. P. King (Rockwell Hanford Operations)	*MAGNUM2D*[a]*–CHAINT*[b] (1981) Multicomponent nuclide transport, basalt repository	Two-dimensional isoparametric elements for porous matrix One-dimensional elements for discrete fractures Galerkin Solute transport in fractures	Nonisothermal Convection-diffusion Double porosity	Advection Dispersion Sorption Decay chains Mass release
J. O. Duguid M. Reeves (Oak Ridge Natl. Lab.)	*Dissolved Constituent Transport Code* (1976) Seepage pond	Two-dimensional vertical Galerkin Linear basis functions L-U decomposition	Given saturated or unsaturated flow	Advection Dispersion Diffusion Adsorption Decay

TABLE 20. (continued)

Authors (Institute)	*Code Name* (Reference Date) Simulation Examples	Numerical Characteristics	Fluid Flow Processes	Solute Transport Processes
D. B. Grove (U. S. Geological Survey, Denver)	*Grove/Galerkin* (1977) Chloride, Tritium, and ^{90}Sr transport	Two-dimensional areal Galerkin Linear and Hermite cubic basis functions SOR iterative	Finite difference for flow	Advection Dispersion Decay Ion exchange
S. K. Gupta C. R. Cole C. T. Kincaid F. E. Kaszeta (Pacific Northwest Lab.)	*FE3DGW, CFEST*[c] (1980) Hypothetical salt and hard rock repositories	Three-dimensional Galerkin *FE3DGW*-mixed order isoparametric elements *CFEST*-linear elements Sequential solution	Transient Multi-aquifer Nonisothermal Convection	Advection Dispersion Diffusion
P. S. Huyakorn B. H. Lester J. W. Mercer (GeoTrans, Inc.)	*FTRANS*[d] (1983) Transport of ^{237}Np from a waste repository in a uniform flow field	Two-dimensional for flow and transport in fractures One-dimensional for the matrix Upstream weighting for fractures	Transient Flow through fractured porous media	Advection[e] Dispersion[e] Diffusion[f] Adsorption[f] Decay
Intera Environmental Consultants, Inc.	*VCHFLD* Multi-species chemical transport	Two-dimensional areal or cross-section Galerkin Bilinear basis functions Gaussian elimination or SOR	Confined aquifer Injection/ production	Advection Sorption Multi-species reactions
J. Marlon-Lambert I. Miller (Golder Assoc.)	*Groundwater computer package* (1978)	Two-dimensional areal or axisymmetric Galerkin L-U decomposition, Doolittle method	Transient Layered aquifer Confined or unconfined aquifer	Advection Dispersion Diffusion Adsorption Decay Io n exchange

TABLE 20. (continued)

Authors (Institute)	*Code Name* (Reference Date) Simulation Examples	Numerical Characteristics	Fluid Flow Processes	Solute Transport Processes
J. Noorishad M. Mehran (Lawrence Berkeley Lab.)	*Flows* (1982)	Two-dimensional areal or cross-section Galerkin Upstream weighting Quadrilateral isoparametric elements	Transient Discrete fractures	Advection Dispersion Adsorption Decay
J. F. Pickens G. E. Grisak (Environment Canada)	*SHALT* (1979) Well testing, Chalk River	Two-dimensional areal or cross-section Triangular elements Sequential solution	Transient Nonisothermal Convection	Advection Dispersion Sorption; Decay
L. Picking (Stone and Webster Engineering)	*SALTRP* (*Frind and Trudeau*, 1980) Salt transport and dissolution	Two-dimensional vertical cross-section Galerkin	Transient Confined aquifer	Advection Dispersion Diffusion Dissolution
G. F. Pinder (Princeton Univ.)	*ISOQUAD* (1973) Chromium contamination, Long Island, N. Y.	Two-dimensional areal Galerkin	Transient Confined aquifer	Advection Dispersion Diffusion
G. Segol E. O. Frind (Univ. Waterloo)	*3D Saturated/ Unsaturated Transport Model* (1976) Flow from ponds	Three-dimensional Galerkin Isoparametric elements	Transient Saturated/ unsaturated Free surface	Advection Dispersion Diffusion Adsorption Decay
J. W. Warner (Colorado State Univ.)	*RESTOR* (1981) Two solute transport	Two-dimensional areal Triangular elements Gauss-Seidel or point SOR Leap frog solution	Transient Confined or leaky aquifer	Advection Dispersion Diffusion Ion exchange

TABLE 20. (continued)

Authors (Institute)	*Code Name* (Reference Date) Simulation Examples	Numerical Characteristics	Fluid Flow Processes	Solute Transport Processes
G. T. Yeh D. S. Ward (Oak Ridge Natl. Lab.)	*FEMWATER*[a]–*FEMWASTE*[b] (1981) Seepage pond	Two-dimensional areal or cross-section Upstream weighting Quadrilateral bilinear elements	Saturated/ unsaturated Density as function of moisture content	Advection Dispersion Sorption; Decay

[a]Calculates fluid flow

[b]Calculates solute transport using fluid flow calculated by (a)

[c]CFEST is an extension of FE3DGW to include energy and solute transport

[d]One member of a family of codes, other members consider variably-saturated flow, solute and energy transport, and nuclide decay chain transport [*International Ground Water Modeling Center*, 1983]

[e]In fractures

[f]In matrix

TABLE 21. Flow path network solute transport codes

Authors (Institute)	*Code Name* (Reference Date) Simulation Examples	Numerical Characteristics	Fluid Flow Processes	Solute Transport Processes
S. W. Ahlstrom H. P. Foote R. C. Arnett C. R. Cole R. J. Serne (Pacific Northwest Lab.)	*MMT* (1977) Radionuclide contamination, Hanford, WA.	Two-dimensional Monte Carlo for dispersion (discrete-parcel-random walk approach)	Given saturated or unsaturated flow	Advection Dispersion Adsorption Decay Ion exchange Precipitation Dissolution

TABLE 21. (continued)

Authors (Institute)	*Code Name* (Reference Date) Simulation Examples	Numerical Characteristics	Fluid Flow Processes	Solute Transport Processes
J. E. Campbell D. E. Longsine R. W. Cranwell (Sandia Natl. Lab.)	*NWFT/DVM* (1981) Radionuclide transport with chain decays	Flow in network of one-dimensional path segments Dispersion by distributed velocity method with Gaussian distribution for contaminant packets	Steady state Stratified sedimentary rocks Density varies with salt concentration	Advection Dispersion Sorption Decay chains Equilibrium solubility Kinetic leaching
R. W. Nelson J. A. Schur (Pacific Northwest Lab.)	*PATHS* *(1980)* Accidental contaminant release evaluation, S.C., WA. Copper tailing seepage	Two-dimensional areal Analytical flow potential Numerical pathline	Steady/transient Uniform confined aquifer Line-source wells Constant head pond	Advection Adsorption Ion exchange
T. A. Prickett T. G. Naymik C. G. Lonnquist (Illinois State Water Survey)	*Random-Walk* (1981)	Two-dimensional areal Particle-in-a-cell for advection Random-walk for dispersion	Transient Confined, semi-confined, or unconfined aquifer	Advection Dispersion Diffusion Adsorption Decay
B. Ross C. M. Koplik M. S. Giuffre S. P. Hodgin J. J. Duffy (Analytic Sciences Co.)	*NUTRAN* (1979) Long-term hazard from waste repositories	Flow in network of one-dimensional paths Green's function for transport	Resaturation of repository Withdrawal through wells	Advection Dispersion Sorption Decay chains Leaching Dissolution Diffusion through barriers

TABLE 21. (continued)

Authors (Institute)	*Code Name* (Reference Date) Simulation Examples	Numerical Characteristics	Fluid Flow Processes	Solute Transport Processes
F. W. Schwartz A. Crowe (Univ. Alberta)	*DPCT* (1980) Long-term effects of waste repositories	Two-dimensional vertical cross-section Particle tracking with random number for longitudinal and transverse dispersion	Steady state Describe water table Finite element for flow	Advection Dispersion Sorption Decay

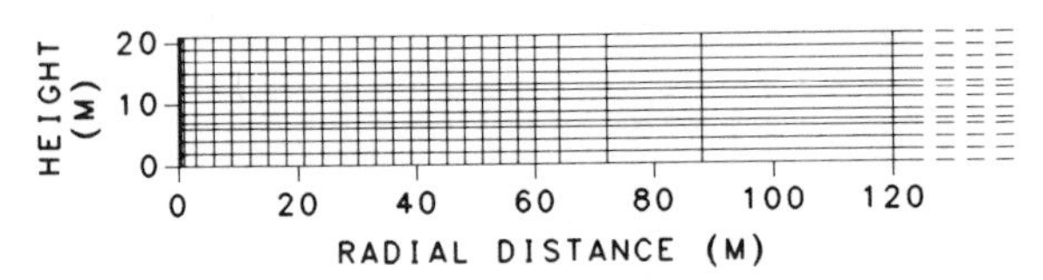

Fig. 28. A vertical cross section of the axisymmetric mesh used for the PT calculation of aquifer temperature. The mesh extends radially beyond 120 m, with increasing width elements, to a distance of 16 km. The first column of elements, which acts as a well bore, does not show because it is only 10 cm wide.

In the present application of the numerical model PT, the control of the shape and movement of a fluid plume in a confined aquifer is studied. We assume a hot water plume with lower density and viscosity than the native groundwater. Various injection and production schemes are investigated to determine their effects on the transformation of the plume and to maximize the withdrawal of the plume fluid from the aquifer.

In any numerical model application, four types of input are needed: (1) fluid properties, (2) material properties, (3) boundary and initial conditions, operation schemes, and sources and sinks, and (4) mesh design. Fluid properties include physical properties, such as the density and viscosity of the contaminant plume and the groundwater. If the properties are similar to those of pure water, the values may be found from handbooks and tables; otherwise they have to be measured. Material properties include aquifer permeability, porosity, etc. The properties used in our application are found in Appendix J, Table J1. A three-layered aquifer is considered, with anisotropic permeability; the average horizontal permeability is seven times greater than the average vertical permeability.

The sources and sinks and boundary and initial conditions must be specified. In many cases, reasonable guesses are made based on geological and hydrological knowledge of the site. Based on these conditions, a mesh is designed. For the present example, a vertical, axisymmetric mesh, shown in Figure 28, is used. The mesh is finer where greater variation in flow is expected.

The hot water plume is created by pumping 80°C water into the injection well at $r = 0$; in all, 42,500 m^3 of hot water are injected at 200 gpm (0.012 m^3/s). The resulting hot water plume in the aquifer is shown in Appendix J, Figure J1. The figure shows that a large buoyancy force has caused much of the hot water to rise to the upper part of the aquifer. Note also the preferential flow into the middle layer of the aquifer due to the high permeability there. The thermal conduction within the aquifer results in a diffuse thermal front, which is evident in the figure.

This combination of the effects of natural convection due to buoyancy, forced convection due to injection, and conduction makes the use of a simple model inadequate; a sophisticated numerical model, such as PT, becomes necessary. A complete report on the use of PT, including an investigation of the various strategies for withdrawal of the fluid plume and validation of the calculational results against field data, is given in Appendix J. The report was prepared by C. Doughty and C. F. Tsang and was originally presented at the ASME-JSME Thermal Engineering Joint Conference held in Honolulu, Hawaii on March 20–24, 1983.

4.6. Advantages of Numerical Methods

Some of the advantages of applying numerical methods for the study of solute transport in the subsurface are listed below.

1. Spatial and temporal variation of system properties such as hydraulic con-

ductivity, porosity, and dispersivity can be easily managed with numerical methods.

2. Field problems with complex boundary conditions are simple to handle with numerical methods.

3. Three dimensional transient problems can be treated without much difficulty.

4.7. Limitations of Numerical Methods

Some of the important limitations of numerical methods are listed below.

1. Application of a complex numerical program requires a certain level of user familiarity with the program. Achievement of such a familiarity is time consuming and could be prohibitive either when dealing with urgent problems or when funding is limited.

2. Very often errors due to numerical dispersion overshadow the physical dispersion of the solute within the porous medium.

3. Preparation of input data for numerical codes often takes a long time. Even if one wishes to solve a simple problem which is manageable with analytical or semianalytical methods, far greater time is needed to prepare the input data for a numerical code.

5 A Discussion on Choice of Methods and Data Needs

The previous three chapters of this report have described methods, at three levels of complication, that can be used to predict the extent of contamination in the subsurface. At each level there are advantages and limitations to be considered for a given situation. In this chapter, we will discuss the kinds of data needed by each method then we shall address the selection of the best method for handling a given problem.

5.1. Data Needs

5.1.1. *Analytical and Semianalytical Methods*

The data needed for these methods are generally simple. The following is a list of data usually required for analytical methods.

1. Geometry of the system (e.g., average thickness and depth of the aquifer), positions of various significant hydrogeologic features (e.g., sources of contamination), and means of groundwater discharge (e.g., production wells and effluent streams).
2. Direction and magnitude of average regional fluid velocity in the vicinity of the study area.
3. Sufficient information about the concentration of different solute species and the rate of leaching or injection, as well as the history of operation at the individual disposal facilities.
4. A representative value of longitudinal dispersivity for one-dimensional problems and both longitudinal and transverse dispersivities for two-dimensional problems.
5. Retardation factor or distribution coefficient for solutes which can be adsorbed onto the media and the radioactive decay factor, if appropriate.

Except for the fourth item, the data mentioned above are also required for semianalytical methods.

Representative thickness and depth of an aquifer are generally obtained from the geological logs of previously drilled wells within the aquifer. Regional velocity can be obtained by using Darcy's law, water table and/or piezometric maps of the region (for the hydraulic gradient), and some average value of hydraulic conductivity, or directly by tracer techniques.

Information about the concentration of different solute species and the rate of injection in the past and present is probably the most difficult part of the data to obtain. Many industries either do not have these data or refuse to give them out.

Finding representative values of dispersivity is not a simple job either. Presently there are two ways to obtain the dispersivity of geological materials. One way is to run a column test on a sample of material in the laboratory and use a breakthrough curve to determine longitudinal dispersivity. Values of dispersivity for relatively homogeneous porous media obtained by this method are generally in the range of 10^{-2} to 1 cm. Another way of obtaining the dispersivity of

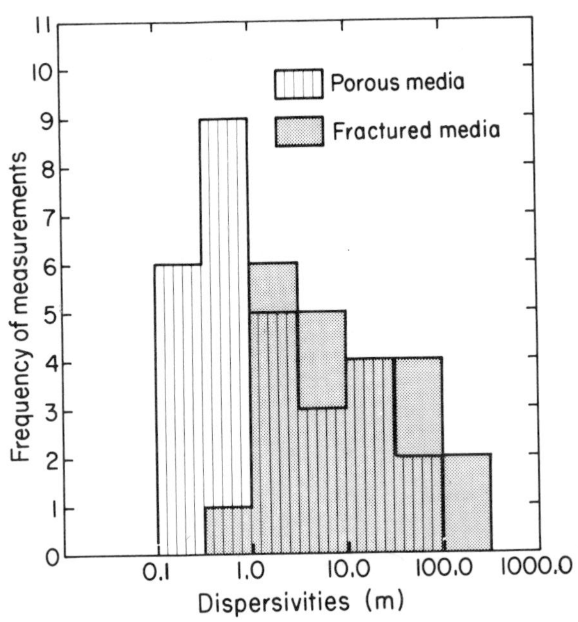

Fig. 29. Distribution of dispersivities of porous and fractured media [modified from *Lallemand-Barres and Peaudecerf*, 1978].

an aquifer in a certain region is by using either single-well or double-well tracer tests. Values of dispersivity obtained by these methods are generally several orders of magnitude larger than the range obtained in the laboratory. Figure 29 presents a distribution of the measured values of dispersivity in both fractured and porous media. Although the general difference in magnitude between laboratory measurements and field calculations of dispersivity is believed to be due to the inhomogeneous and anisotropic nature of groundwater systems, the subject is an open question and requires further study.

For preliminary studies in a case where no dispersivity data are available, one can examine the limits of the range of values of dispersivity which have been previously measured for the same type of geological materials elsewhere [see *Anderson*, 1979]. This enables one to envisage the whole range of possibilities in a very short time and very economically.

5.1.2. *Numerical Methods*

The following is a list of data needed to predict the extent of contamination in a groundwater system by use of a numerical model.

1. Geometry of the system (e.g., areal extent, thickness, and spatial variation of the aquifer, location of natural or mathematical boundaries).
2. Velocity distribution throughout the system.
3. Dispersivity distribution within the system.
4. Present distribution of concentration of various solute species in the system.
5. Complete information about present and future sources of contamination (e.g., rate of injection or leaching, chemical composition, position relative to the aquifer).

6. Location of natural and artificial discharge areas in the system including production wells.

For detailed modeling, the volume of required data is large and acquisition requires a long period of time and an extensive budget. Thus for many cases a complete set of data is not available. Furthermore, such data as velocity and dispersivity distributions are usually obtained through calibration techniques which do not yield unique answers and in turn rely on vast data banks, which do not exist for most situations. Hence considerable hydrologic intuition is necessary to avoid blind acceptance of modeling results which may be in considerable error.

5.2. Selection of Method

In general, there are two types of problems which should be solved by employing solute transport theory [*Anderson*, 1979]: (1) to assess the environmental impact of subsurface waste disposal at a proposed site and (2) to assess the long-term consequences or the effects of remedial measures at an operating site where a contaminant plume has already been detected. The first problem should be solved in two or three different stages. In the preliminary stage of study, one needs a very rough estimate of the probable extent of contamination at some point in the future. This can help determine whether the potential site should be eliminated or kept for further extensive studies. There is often little initial data to work with, suggesting that analytical methods are the most useful tools in the hand of the investigator. As discussed previously, the amount of data required for these methods is relatively small and application of the techniques is fast and simple, so that the effect of uncertainties in the data can be easily evaluated in a short period of time by simply rerunning a problem with different estimates.

If the site involves several sources of contamination and one or more production wells, one has to resort to semianalytical methods. These methods also satisfy the limited data constraints mentioned above.

Once it is established that the site is relatively safe but further study is needed, the problem enters the next stage. The next stage of study requires a detailed site characterization and further application of the analytical and semianalytical methods, based on more reliable data. Application of simple numerical methods may also be advisable at this stage.

Often little or no prior hydrologic or contaminant transport data is available at such proposed sites. Furthermore, the data accumulated during a relatively short period of study are not generally enough to justify application of a complete numerical model. By contrast, at presently operating sites, one would try to obtain the history of operation of the site. It is still advisable to first use semianalytical or analytical methods, then apply numerical simulations. Again, since the amount of available data for application of a numerical contaminant transport model is not usually sufficient, modelers usually proceed by first calibrating the model using historical data and then predict the future extent of contamination with the calibrated model. Such calibration of models is generally necessary because two sets of essential data, velocity and dispersivity distributions throughout the system, are seldom directly available.

The process of calibration includes the following steps.

1. Given the distribution of hydraulic head, the distribution of hydraulic conductivity throughout the system is calculated using an inverse method. The accuracy of the results is a function of the degree of homogeneity of the aquifer and the density of hydraulic head data.

2. By the application of Darcy's law, together with the distribution of hydraulic head and hydraulic conductivity, the distribution of velocity all over the system is calculated.

3. The contaminant transport model is calibrated by adjusting dispersivity and porosity until the model simulates the observed concentrations.

Once the flow and contaminant transport models are calibrated in this way, one can use the values of hydraulic conductivity and dispersivity obtained by calibration to predict the response of the system to future changes such as those resulting from remedial measures.

However, a very important question still remains unanswered. We need to know how we can extend our information beyond the limits of calibration in order to predict the future extent of contamination. One may note that calibration of the contaminant transport model is based on concentration data measured within the plume area. Values of dispersivity beyond the boundaries of the plume are not known. Therefore the model may not be valid for prediction of the extent of contaminant beyond these boundaries.

In general, the application of complicated numerical models requires substantial experience and knowledge of the computer codes. Consultation with model developers and experienced model users is essential. Numerical models are only tools, after all; no special dispensation from conceptual errors automatically accompanies their use.

6 Conclusions

The present handbook introduces the reader to various mathematical methods for estimating solute transport in groundwater systems. We have prepared tables, figures, and simple computer programs that can be directly used for field studies.

Three levels of mathematical methods are covered: (1) analytical, (2) semianalytical, and (3) numerical. The first two levels require relatively small amounts of data. However, representative values of property parameters must be obtained, and results are applicable only for cases where the simplifications required for analytical and semianalytical methods are valid. A number of methods are described; relevant tables and computer programs are presented in Appendices A-I. At the third level, numerical approches are discussed and a number of currently available numerical models are listed, indicating code capabilities and code developers to be contacted for further information. An example of the use of one such model is presented in Appendix J. A discussion on method selection and data requirements concludes the handbook.

In general, the use of mathematical methods requires a certain amount of experience. Use of numerical methods often involves decisions regarding time steps, spatial grid design, upstream weighting factors, or implicit-explicit parameters, as well as the best ways to avoid truncation errors and numerical oscillations. If these decisions are made incorrectly, extraordinary computer effort may be required for a given calculation or wrong answers may be obtained. On the other hand, use of analytical and semianalytical methods, though without these numerical pitfalls, involves decisions concerning the applicability of underlying assumptions and simplifications, as well as the determination of appropriate representative (average) values for material properties. It is hoped that the present work will help readers to begin to obtain the experience and expertise necessary to handle subsurface solute transport problems.

We have not attempted to cover reactive chemical transport in groundwater in the present work. Much research and development continues on this topic and it may be the subject of a future handbook. Comments and suggestions to the authors concerning improvements to this handbook are most welcome.

Appendix A

Tables of Dimensionless Concentration for One-Dimensional Contaminant Transport in Aquifers With One-Dimensional Uniform Flow

The following tables list dimensionless concentration C/C_o (evaluated from (35)) for the following values: velocity v = 0.01–1.0 m/d; dispersion coefficient D = 0.01–50 m^2/d; three values of α_L, 1, 10, and 50 m were used for each value of velocity; retardation factor $R = 1$; decay constant $\lambda = 0$ d^{-1}; decay factor of source α = 0 d^{-1}; period of activity of the source t_o = 60 years; and C/C_o is given as a function of time t in years for different values of distance x in meters.

DIMENSIONLESS CONCENTRATION C/C0 FOR

V = .01 D = .01 R = 1.0 LAMBDA = .0 ALPHA = .0 T0 = 60.0

T(YEARS)	X = 10.	X = 20.	X = 30.	X = 40.	X = 50.	X = 100.
1.0	.7006D–02	.3736D–09	.0000D+00	.0000D+00	.0000D+00	.0000D+00
2.0	.2257D+00	.3413D–03	.8955D–09	.0000D+00	.0000D+00	.0000D+00
3.0	.5778D+00	.2377D–01	.1825D–04	.1834D–09	.0000D+00	.0000D+00
4.0	.8086D+00	.1522D+00	.1906D–02	.1019D–05	.2031D–10	.0000D+00
5.0	.9205D+00	.3817D+00	.2408D–01	.1366D–03	.5843D–07	.0000D+00
6.0	.9684D+00	.6138D+00	.1067D+00	.2840D–02	.9272D–05	.0000D+00
7.0	.9877D+00	.7852D+00	.2632D+00	.2045D–01	.2804D–03	.0000D+00
8.0	.9953D+00	.8901D+00	.4572D+00	.7643D–01	.3020D–02	.0000D+00
9.0	.9982D+00	.9470D+00	.6391D+00	.1861D+00	.1641D–01	.0000D+00
10.0	.9993D+00	.9756D+00	.7798D+00	.3394D+00	.5551D–01	.4338D–13
15.0	.1000D+01	.9996D+00	.9918D+00	.9232D+00	.6772D+00	.7117D–05
20.0	.1000D+01	.1000D+01	.9998D+00	.9971D+00	.9728D+00	.1251D–01
25.0	.1000D+01	.1000D+01	.1000D+01	.9999D+00	.9990D+00	.2589D+00
30.0	.1000D+01	.1000D+01	.1000D+01	.1000D+01	.1000D+01	.7419D+00
35.0	.1000D+01	.1000D+01	.1000D+01	.1000D+01	.1000D+01	.9598D+00
40.0	.1000D+01	.1000D+01	.1000D+01	.1000D+01	.1000D+01	.9966D+00
45.0	.1000D+01	.1000D+01	.1000D+01	.1000D+01	.1000D+01	.9998D+00
50.0	.1000D+01	.1000D+01	.1000D+01	.1000D+01	.1000D+01	.1000D+01
60.0	.1000D+01	.1000D+01	.1000D+01	.1000D+01	.1000D+01	.1000D+01

DIMENSIONLESS CONCENTRATION C/C0 FOR

V = .01 D = .01 R = 1.0 LAMBDA = .0 ALPHA = .0 T0 = 60.0

T(YEARS)	X = 150.	X = 200.	X = 250.	X = 300.	X = 400.	X = 500.
1.0	.0000D+00	.0000D+00	.0000D+00	.0000D+00	.0000D+00	.0000D+00
2.0	.0000D+00	.0000D+00	.0000D+00	.0000D+00	.0000D+00	.0000D+00
3.0	.0000D+00	.0000D+00	.0000D+00	.0000D+00	.0000D+00	.0000D+00
4.0	.0000D+00	.0000D+00	.0000D+00	.0000D+00	.0000D+00	.0000D+00
5.0	.0000D+00	.0000D+00	.0000D+00	.0000D+00	.0000D+00	.0000D+00
6.0	.0000D+00	.0000D+00	.0000D+00	.0000D+00	.0000D+00	.0000D+00
7.0	.0000D+00	.0000D+00	.0000D+00	.0000D+00	.0000D+00	.0000D+00
8.0	.0000D+00	.0000D+00	.0000D+00	.0000D+00	.0000D+00	.0000D+00
9.0	.0000D+00	.0000D+00	.0000D+00	.0000D+00	.0000D+00	.0000D+00
10.0	.0000D+00	.0000D+00	.0000D+00	.0000D+00	.0000D+00	.0000D+00
15.0	.0000D+00	.0000D+00	.0000D+00	.0000D+00	.0000D+00	.0000D+00
20.0	.8496D–10	.0000D+00	.0000D+00	.0000D+00	.0000D+00	.0000D+00
25.0	.6592D–05	.0000D+00	.0000D+00	.0000D+00	.0000D+00	.0000D+00
30.0	.3072D–02	.4592D–09	.0000D+00	.0000D+00	.0000D+00	.0000D+00
35.0	.8212D–01	.3031D–05	.0000D+00	.0000D+00	.0000D+00	.0000D+00
40.0	.4093D+00	.7858D–03	.5644D–09	.0000D+00	.0000D+00	.0000D+00
45.0	.7865D+00	.2438D–01	.1108D–05	.3364D–13	.0000D+00	.0000D+00
50.0	.9566D+00	.1810D+00	.2059D–03	.3835D–09	.0000D+00	.0000D+00
60.0	.9995D+00	.8204D+00	.6988D–01	.5482D–04	.0000D+00	.0000D+00

DIMENSIONLESS CONCENTRATION C/C0 FOR

V = .01 D = .10 R = 1.0 LAMBDA = .0 ALPHA = .0 T0 = 60.0

T(YEARS)	X = 10.	X = 20.	X = 30.	X = 40.	X = 50.	X = 100.
1.0	.1334D+00	.1238D-01	.3586D-03	.3039D-05	.7125D-08	.0000D+00
2.0	.3172D+00	.9541D-01	.1668D-01	.1617D-02	.8549D-04	.0000D+00
3.0	.4551D+00	.2083D+00	.6758D-01	.1506D-01	.2253D-02	.3106D-09
4.0	.5585D+00	.3181D+00	.1417D+00	.4813D-01	.1224D-01	.1258D-06
5.0	.6378D+00	.4157D+00	.2246D+00	.9844D-01	.3448D-01	.4743D-05
6.0	.6999D+00	.4998D+00	.3074D+00	.1599D+00	.6934D-01	.5398D-04
7.0	.7493D+00	.5713D+00	.3855D+00	.2266D+00	.1145D+00	.3077D-03
8.0	.7892D+00	.6319D+00	.4571D+00	.2944D+00	.1669D+00	.1135D-02
9.0	.8218D+00	.6834D+00	.5216D+00	.3606D+00	.2233D+00	.3126D-02
10.0	.8485D+00	.7270D+00	.5789D+00	.4233D+00	.2814D+00	.7006D-02
15.0	.9292D+00	.8660D+00	.7783D+00	.6690D+00	.5459D+00	.7486D-01
20.0	.9648D+00	.9315D+00	.8819D+00	.8143D+00	.7291D+00	.2257D+00
25.0	.9818D+00	.9640D+00	.9361D+00	.8959D+00	.8417D+00	.4092D+00
30.0	.9904D+00	.9806D+00	.9649D+00	.9414D+00	.9081D+00	.5778D+00
35.0	.9948D+00	.9894D+00	.9805D+00	.9668D+00	.9466D+00	.7114D+00
40.0	.9971D+00	.9941D+00	.9890D+00	.9810D+00	.9690D+00	.8086D+00
45.0	.9984D+00	.9967D+00	.9938D+00	.9891D+00	.9819D+00	.8757D+00
50.0	.9991D+00	.9981D+00	.9964D+00	.9937D+00	.9894D+00	.9205D+00
60.0	.9997D+00	.9994D+00	.9988D+00	.9979D+00	.9963D+00	.9684D+00

DIMENSIONLESS CONCENTRATION C/C0 FOR

V = .01 D = .10 R = 1.0 LAMBDA = .0 ALPHA = .0 T0 = 60.0

T(YEARS)	X = 150.	X = 200.	X = 250.	X = 300.	X = 400.	X = 500.
1.0	.0000D+00	.0000D+00	.0000D+00	.0000D+00	.0000D+00	.0000D+00
2.0	.0000D+00	.0000D+00	.0000D+00	.0000D+00	.0000D+00	.0000D+00
3.0	.0000D+00	.0000D+00	.0000D+00	.0000D+00	.0000D+00	.0000D+00
4.0	.0000D+00	.0000D+00	.0000D+00	.0000D+00	.0000D+00	.0000D+00
5.0	.1032D-11	.0000D+00	.0000D+00	.0000D+00	.0000D+00	.0000D+00
6.0	.2054D-09	.0000D+00	.0000D+00	.0000D+00	.0000D+00	.0000D+00
7.0	.9048D-08	.0000D+00	.0000D+00	.0000D+00	.0000D+00	.0000D+00
8.0	.1548D-06	.3516D-12	.0000D+00	.0000D+00	.0000D+00	.0000D+00
9.0	.1406D-05	.1691D-10	.0000D+00	.0000D+00	.0000D+00	.0000D+00
10.0	.8193D-05	.3736D-09	.0000D+00	.0000D+00	.0000D+00	.0000D+00
15.0	.1519D-02	.3780D-05	.1068D-08	.3304D-13	.0000D+00	.0000D+00
20.0	.1873D-01	.3413D-03	.1255D-05	.8955D-09	.0000D+00	.0000D+00
25.0	.7722D-01	.4592D-02	.7835D-04	.3662D-06	.1515D-12	.0000D+00
30.0	.1841D+00	.2377D-01	.1119D-02	.1825D-04	.1834D-09	.0000D+00
35.0	.3221D+00	.7126D-01	.6867D-02	.2720D-03	.2657D-07	.5892D-13
40.0	.4663D+00	.1522D+00	.2487D-01	.1906D-02	.1019D-05	.2031D-10
45.0	.5979D+00	.2601D+00	.6347D-01	.8074D-02	.1611D-04	.1766D-08
50.0	.7075D+00	.3817D+00	.1270D+00	.2408D-01	.1366D-03	.5843D-07
60.0	.8567D+00	.6138D+00	.3157D+00	.1067D+00	.2840D-02	.9272D-05

DIMENSIONLESS CONCENTRATION C/C0 FOR

V = .01 D = .50 R = 1.0 LAMBDA = .0 ALPHA = .0 T0 = 60.0

T(YEARS)	X = 10.	X = 20.	X = 30.	X = 40.	X = 50.	X = 100.
1.0	.1449D+00	.6476D–01	.2367D–01	.6969D–02	.1634D–02	.2963D–07
2.0	.2458D+00	.1522D+00	.8564D–01	.4354D–01	.1988D–01	.6911D–04
3.0	.3191D+00	.2245D+00	.1485D+00	.9194D–01	.5315D–01	.1118D–02
4.0	.3772D+00	.2849D+00	.2056D+00	.1415D+00	.9253D–01	.4885D–02
5.0	.4253D+00	.3365D+00	.2569D+00	.1888D+00	.1334D+00	.1238D–01
6.0	.4664D+00	.3814D+00	.3028D+00	.2331D+00	.1737D+00	.2363D–01
7.0	.5021D+00	.4209D+00	.3442D+00	.2742D+00	.2125D+00	.3818D–01
8.0	.5337D+00	.4562D+00	.3816D+00	.3122D+00	.2494D+00	.5540D–01
9.0	.5618D+00	.4879D+00	.4158D+00	.3474D+00	.2843D+00	.7466D–01
10.0	.5872D+00	.5166D+00	.4470D+00	.3800D+00	.3172D+00	.9541D–01
15.0	.6845D+00	.6282D+00	.5706D+00	.5125D+00	.4551D+00	.2083D+00
20.0	.7509D+00	.7054D+00	.6578D+00	.6086D+00	.5585D+00	.3181D+00
25.0	.7993D+00	.7620D+00	.7225D+00	.6809D+00	.6378D+00	.4157D+00
30.0	.8358D+00	.8050D+00	.7720D+00	.7368D+00	.6999D+00	.4998D+00
35.0	.8643D+00	.8385D+00	.8107D+00	.7810D+00	.7493D+00	.5713D+00
40.0	.8868D+00	.8652D+00	.8417D+00	.8163D+00	.7892D+00	.6319D+00
45.0	.9049D+00	.8866D+00	.8667D+00	.8451D+00	.8218D+00	.6834D+00
50.0	.9196D+00	.9041D+00	.8871D+00	.8686D+00	.8485D+00	.7270D+00
60.0	.9418D+00	.9305D+00	.9180D+00	.9043D+00	.8893D+00	.7958D+00

DIMENSIONLESS CONCENTRATION C/C0 FOR

V = .01 D = .50 R = 1.0 LAMBDA = .0 ALPHA = .0 T0 = 60.0

T(YEARS)	X = 150.	X = 200.	X = 250.	X = 300.	X = 400.	X = 500.
1.0	.0000D+00	.0000D+00	.0000D+00	.0000D+00	.0000D+00	.0000D+00
2.0	.1089D–07	.6637D–13	.0000D+00	.0000D+00	.0000D+00	.0000D+00
3.0	.3159D–05	.1060D–08	.4020D–13	.0000D+00	.0000D+00	.0000D+00
4.0	.5900D–04	.1482D–06	.7388D–10	.0000D+00	.0000D+00	.0000D+00
5.0	.3586D–03	.3039D–05	.7125D–08	.4522D–11	.0000D+00	.0000D+00
6.0	.1239D–02	.2360D–04	.1555D–06	.3477D–09	.0000D+00	.0000D+00
7.0	.3067D–02	.1043D–03	.1441D–05	.7934D–08	.1443D–13	.0000D+00
8.0	.6142D–02	.3210D–03	.7802D–05	.8437D–07	.8535D–12	.0000D+00
9.0	.1065D–01	.7845D–03	.2934D–04	.5377D–06	.2069D–10	.0000D+00
10.0	.1668D–01	.1617D–02	.8549D–04	.2390D–05	.2679D–09	.0000D+00
15.0	.6758D–01	.1506D–01	.2253D–02	.2256D–03	.6290D–06	.3106D–09
20.0	.1417D+00	.4813D–01	.1224D–01	.2314D–02	.3239D–04	.1258D–06
25.0	.2246D+00	.9844D–01	.3448D–01	.9531D–02	.3533D–03	.4743D–05
30.0	.3074D+00	.1599D+00	.6934D–01	.2482D–01	.1756D–02	.5398D–04
35.0	.3855D+00	.2266D+00	.1145D+00	.4931D–01	.5541D–02	.3077D–03
40.0	.4571D+00	.2944D+00	.1669D+00	.8250D–01	.1311D–01	.1135D–02
45.0	.5216D+00	.3606D+00	.2233D+00	.1229D+00	.2558D–01	.3126D–02
50.0	.5789D+00	.4233D+00	.2814D+00	.1686D+00	.4352D–01	.7006D–02
60.0	.6744D+00	.5356D+00	.3951D+00	.2688D+00	.9525D–01	.2324D–01

DIMENSIONLESS CONCENTRATION C/C0 FOR

V = .02 D = .02 R = 1.0 LAMBDA =.0 ALPHA = .0 T0 = 60.0

T(YEARS)	X = 10.	X = 20.	X = 30.	X = 40.	X = 50.	X = 100.
1.0	.2257D+00	.3413D–03	.8955D–09	.0000D+00	.0000D+00	.0000D+00
2.0	.8086D+00	.1522D+00	.1906D–02	.1019D–05	.2031D–10	.0000D+00
3.0	.9684D+00	.6138D+00	.1067D+00	.2840D–02	.9272D–05	.0000D+00
4.0	.9953D+00	.8901D+00	.4572D+00	.7643D–01	.3020D–02	.0000D+00
5.0	.9993D+00	.9756D+00	.7798D+00	.3394D+00	.5551D–01	.4338D–13
6.0	.9999D+00	.9953D+00	.9327D+00	.6596D+00	.2523D+00	.8342D–09
7.0	.1000D+01	.9991D+00	.9829D+00	.8669D+00	.5444D+00	.6015D–06
8.0	.1000D+01	.9999D+00	.9962D+00	.9576D+00	.7842D+00	.5599D–04
9.0	.1000D+01	.1000D+01	.9992D+00	.9884D+00	.9169D+00	.1340D–02
10.0	.1000D+01	.1000D+01	.9998D+00	.9971D+00	.9728D+00	.1251D–01
15.0	.1000D+01	.1000D+01	.1000D+01	.1000D+01	.1000D+01	.7419D+00
20.0	.1000D+01	.1000D+01	.1000D+01	.1000D+01	.1000D+01	.9966D+00
25.0	.1000D+01	.1000D+01	.1000D+01	.1000D+01	.1000D+01	.1000D+01
30.0	.1000D+01	.1000D+01	.1000D+01	.1000D+01	.1000D+01	.1000D+01
35.0	.1000D+01	.1000D+01	.1000D+01	.1000D+01	.1000D+01	.1000D+01
40.0	.1000D+01	.1000D+01	.1000D+01	.1000D+01	.1000D+01	.1000D+01
45.0	.1000D+01	.1000D+01	.1000D+01	.1000D+01	.1000D+01	.1000D+01
50.0	.1000D+01	.1000D+01	.1000D+01	.1000D+01	.1000D+01	.1000D+01
60.0	.1000D+01	.1000D+01	.1000D+01	.1000D+01	.1000D+01	.1000D+01

DIMENSIONLESS CONCENTRATION C/C0 FOR

V = .02 D = .02 R = 1.0 LAMBDA = .0 ALPHA = .0 T0 = 60.0

T(YEARS)	X = 150.	X = 200.	X = 250.	X = 300.	X = 400.	X = 500.
1.0	.0000D+00	.0000D+00	.0000D+00	.0000D+00	.0000D+00	.0000D+00
2.0	.0000D+00	.0000D+00	.0000D+00	.0000D+00	.0000D+00	.0000D+00
3.0	.0000D+00	.0000D+00	.0000D+00	.0000D+00	.0000D+00	.0000D+00
4.0	.0000D+00	.0000D+00	.0000D+00	.0000D+00	.0000D+00	.0000D+00
5.0	.0000D+00	.0000D+00	.0000D+00	.0000D+00	.0000D+00	.0000D+00
6.0	.0000D+00	.0000D+00	.0000D+00	.0000D+00	.0000D+00	.0000D+00
7.0	.0000D+00	.0000D+00	.0000D+00	.0000D+00	.0000D+00	.0000D+00
8.0	.0000D+00	.0000D+00	.0000D+00	.0000D+00	.0000D+00	.0000D+00
9.0	.8510D–13	.0000D+00	.0000D+00	.0000D+00	.0000D+00	.0000D+00
10.0	.8496D–10	.0000D+00	.0000D+00	.0000D+00	.0000D+00	.0000D+00
15.0	.3072D–02	.4592D–09	.0000D+00	.0000D+00	.0000D+00	.0000D+00
20.0	.4093D+00	.7858D–03	.5644D–09	.0000D+00	.0000D+00	.0000D+00
25.0	.9566D+00	.1810D+00	.2059D–03	.3835D–09	.0000D+00	.0000D+00
30.0	.9995D+00	.8204D+00	.6988D–01	.5482D–04	.0000D+00	.0000D+00
35.0	.1000D+01	.9932D+00	.5992D+00	.2479D–01	.8283D–10	.0000D+00
40.0	.1000D+01	.9999D+00	.9598D+00	.3732D+00	.4006D–05	.0000D+00
45.0	.1000D+01	.1000D+01	.9990D+00	.8691D+00	.2691D–02	.1144D–10
50.0	.1000D+01	.1000D+01	.1000D+01	.9922D+00	.9898D–01	.3010D–06
60.0	.1000D+01	.1000D+01	.1000D+01	.1000D+01	.9024D+00	.1849D–01

DIMENSIONLESS CONCENTRATION C/C0 FOR

V = .02 D = .20 R = 1.0 LAMBDA = .0 ALPHA = .0 T0 = 60.0

T(YEARS)	X = 10.	X = 20.	X = 30.	X = 40.	X = 50.	X = 100.
1.0	.3172D+00	.9541D–01	.1668D–01	.1617D–02	.8549D–04	.0000D+00
2.0	.5585D+00	.3181D+00	.1417D+00	.4813D–01	.1224D–01	.1258D–06
3.0	.6999D+00	.4998D+00	.3074D+00	.1599D+00	.6934D–01	.5398D–04
4.0	.7892D+00	.6319D+00	.4571D+00	.2944D+00	.1669D+00	.1135D–02
5.0	.8485D+00	.7270D+00	.5789D+00	.4233D+00	.2814D+00	.7006D–02
6.0	.8893D+00	.7958D+00	.6744D+00	.5356D+00	.3951D+00	.2324D–01
7.0	.9180D+00	.8461D+00	.7481D+00	.6290D+00	.4990D+00	.5383D–01
8.0	.9387D+00	.8832D+00	.8048D+00	.7049D+00	.5893D+00	.9949D–01
9.0	.9537D+00	.9108D+00	.8484D+00	.7658D+00	.6657D+00	.1580D+00
10.0	.9648D+00	.9315D+00	.8819D+00	.8143D+00	.7291D+00	.2257D+00
15.0	.9904D+00	.9806D+00	.9649D+00	.9414D+00	.9081D+00	.5778D+00
20.0	.9971D+00	.9941D+00	.9890D+00	.9810D+00	.9690D+00	.8086D+00
25.0	.9991D+00	.9981D+00	.9964D+00	.9937D+00	.9894D+00	.9205D+00
30.0	.9997D+00	.9994D+00	.9988D+00	.9979D+00	.9963D+00	.9684D+00
35.0	.9999D+00	.9998D+00	.9996D+00	.9993D+00	.9987D+00	.9877D+00
40.0	.1000D+01	.9999D+00	.9999D+00	.9997D+00	.9996D+00	.9953D+00
45.0	.1000D+01	.1000D+01	.1000D+01	.9999D+00	.9998D+00	.9982D+00
50.0	.1000D+01	.1000D+01	.1000D+01	.1000D+01	.9999D+00	.9993D+00
60.0	.1000D+01	.1000D+01	.1000D+01	.1000D+01	.1000D+01	.9999D+00

DIMENSIONLESS CONCENTRATION C/C0 FOR

V = .02 D = .20 R = 1.0 LAMBDA = .0 ALPHA = .0 T0 = 60.0

T(YEARS)	X = 150.	X = 200.	X = 250.	X = 300.	X = 400.	X = 500.
1.0	.0000D+00	.0000D+00	.0000D+00	.0000D+00	.0000D+00	.0000D+00
2.0	.0000D+00	.0000D+00	.0000D+00	.0000D+00	.0000D+00	.0000D+00
3.0	.2054D–09	.0000D+00	.0000D+00	.0000D+00	.0000D+00	.0000D+00
4.0	.1548D–06	.3516D–12	.0000D+00	.0000D+00	.0000D+00	.0000D+00
5.0	.8193D–05	.3736D–09	.0000D+00	.0000D+00	.0000D+00	.0000D+00
6.0	.1135D–03	.3826D–07	.8280D–12	.0000D+00	.0000D+00	.0000D+00
7.0	.7279D–03	.1024D–05	.1388D–09	.0000D+00	.0000D+00	.0000D+00
8.0	.2878D–02	.1180D–04	.6333D–08	.4293D–12	.0000D+00	.0000D+00
9.0	.8227D–02	.7735D–04	.1210D–06	.3033D–10	.0000D+00	.0000D+00
10.0	.1873D–01	.3413D–03	.1255D–05	.8955D–09	.0000D+00	.0000D+00
15.0	.1841D+00	.2377D–01	.1119D–02	.1825D–04	.1834D–09	.0000D+00
20.0	.4663D+00	.1522D+00	.2487D–01	.1906D–02	.1019D–05	.2031D–10
25.0	.7075D+00	.3817D+00	.1270D+00	.2408D–01	.1366D–03	.5843D–07
30.0	.8567D+00	.6138D+00	.3157D+00	.1067D+00	.2840D–02	.9272D–05
35.0	.9347D+00	.7852D+00	.5302D+00	.2632D+00	.2045D–01	.2804D–03
40.0	.9716D+00	.8901D+00	.7114D+00	.4572D+00	.7643D–01	.3020D–02
45.0	.9880D+00	.9470D+00	.8375D+00	.6391D+00	.1861D+00	.1641D–01
50.0	.9950D+00	.9756D+00	.9144D+00	.7798D+00	.3394D+00	.5551D–01
60.0	.9992D+00	.9953D+00	.9795D+00	.9327D+00	.6596D+00	.2523D+00

DIMENSIONLESS CONCENTRATION C/C0 FOR

V = .02 D = 1.00 R = 1.0 LAMBDA = .0 ALPHA = .0 T0 = 60.0

T(YEARS)	X = 10.	X = 20.	X = 30.	X = 40.	X = 50.	X = 100.
1.0	.2458D+00	.1522D+00	.8564D–01	.4354D–01	.1988D–01	.6911D–04
2.0	.3772D+00	.2849D+00	.2056D+00	.1415D+00	.9253D–01	.4885D–02
3.0	.4664D+00	.3814D+00	.3028D+00	.2331D+00	.1737D+00	.2363D–01
4.0	.5337D+00	.4562D+00	.3816D+00	.3122D+00	.2494D+00	.5540D–01
5.0	.5872D+00	.5166D+00	.4470D+00	.3800D+00	.3172D+00	.9541D–01
6.0	.6311D+00	.5668D+00	.5022D+00	.4386D+00	.3773D+00	.1396D+00
7.0	.6681D+00	.6093D+00	.5494D+00	.4895D+00	.4307D+00	.1854D+00
8.0	.6997D+00	.6458D+00	.5904D+00	.5342D+00	.4782D+00	.2310D+00
9.0	.7270D+00	.6776D+00	.6262D+00	.5736D+00	.5205D+00	.2754D+00
10.0	.7509D+00	.7054D+00	.6578D+00	.6086D+00	.5585D+00	.3181D+00
15.0	.8358D+00	.8050D+00	.7720D+00	.7368D+00	.6999D+00	.4998D+00
20.0	.8868D+00	.8652D+00	.8417D+00	.8163D+00	.7892D+00	.6319D+00
25.0	.9196D+00	.9041D+00	.8871D+00	.8686D+00	.8485D+00	.7270D+00
30.0	.9418D+00	.9305D+00	.9180D+00	.9043D+00	.8893D+00	.7958D+00
35.0	.9572D+00	.9488D+00	.9395D+00	.9293D+00	.9180D+00	.8461D+00
40.0	.9681D+00	.9619D+00	.9549D+00	.9472D+00	.9387D+00	.8832D+00
45.0	.9761D+00	.9714D+00	.9661D+00	.9602D+00	.9537D+00	.9108D+00
50.0	.9819D+00	.9783D+00	.9743D+00	.9698D+00	.9648D+00	.9315D+00
60.0	.9894D+00	.9873D+00	.9849D+00	.9823D+00	.9793D+00	.9591D+00

DIMENSIONLESS CONCENTRATION C/C0 FOR

V = .02 D = 1.00 R = 1.0 LAMBDA = .0 ALPHA = .0 T0 = 60.0

T(YEARS)	X = 150.	X = 200.	X = 250.	X = 300.	X = 400.	X = 500.
1.0	.1089D–07	.6637D–13	.0000D+00	.0000D+00	.0000D+00	.0000D+00
2.0	.5900D–04	.1482D–06	.7388D–10	.0000D+00	.0000D+00	.0000D+00
3.0	.1239D–02	.2360D–04	.1555D–06	.3477D–09	.0000D+00	.0000D+00
4.0	.6142D–02	.3210D–03	.7802D–05	.8437D–07	.8535D–12	.0000D+00
5.0	.1668D–01	.1617D–02	.8549D–04	.2390D–05	.2679D–09	.0000D+00
6.0	.3319D–01	.4872D–02	.4336D–03	.2288D–04	.1277D–07	.8031D–12
7.0	.5501D–01	.1088D–01	.1399D–02	.1169D–03	.2058D–06	.5637D–10
8.0	.8108D–01	.2007D–01	.3427D–02	.4021D–03	.1677D–05	.1387D–08
9.0	.1103D+00	.3254D–01	.6931D–02	.1060D–02	.8665D–05	.1690D–07
10.0	.1417D+00	.4813D–01	.1224D–01	.2314D–02	.3239D–04	.1258D–06
15.0	.3074D+00	.1599D+00	.6934D–01	.2482D–01	.1756D–02	.5398D–04
20.0	.4571D+00	.2944D+00	.1669D+00	.8250D–01	.1311D–01	.1135D–02
25.0	.5789D+00	.4233D+00	.2814D+00	.1686D+00	.4352D–01	.7006D–02
30.0	.6744D+00	.5356D+00	.3951D+00	.2688D+00	.9525D–01	.2324D–01
35.0	.7481D+00	.6290D+00	.4990D+00	.3709D+00	.1649D+00	.5383D–01
40.0	.8048D+00	.7049D+00	.5893D+00	.4672D+00	.2455D+00	.9949D–01
45.0	.8484D+00	.7658D+00	.6657D+00	.5540D+00	.3304D+00	.1580D+00
50.0	.8819D+00	.8143D+00	.7291D+00	.6298D+00	.4143D+00	.2257D+00
60.0	.9279D+00	.8832D+00	.8236D+00	.7492D+00	.5666D+00	.3726D+00

DIMENSIONLESS CONCENTRATION C/C0 FOR

V = .04 D = .04 R = 1.0 LAMBDA = .0 ALPHA = .0 T0 = 60.0

T(YEARS)	X = 50.	X = 100.	X = 150.	X = 200.	X = 250.	X = 300.
1.0	.2031D–10	.0000D+00	.0000D+00	.0000D+00	.0000D+00	.0000D+00
2.0	.3020D–02	.0000D+00	.0000D+00	.0000D+00	.0000D+00	.0000D+00
3.0	.2523D+00	.8342D–09	.0000D+00	.0000D+00	.0000D+00	.0000D+00
4.0	.7842D+00	.5599D–04	.0000D+00	.0000D+00	.0000D+00	.0000D+00
5.0	.9728D+00	.1251D–01	.8496D–10	.0000D+00	.0000D+00	.0000D+00
6.0	.9979D+00	.1743D+00	.1159D–05	.0000D+00	.0000D+00	.0000D+00
7.0	.9999D+00	.5631D+00	.4062D–03	.3700D–11	.0000D+00	.0000D+00
8.0	.1000D+01	.8662D+00	.1486D–01	.2522D–07	.0000D+00	.0000D+00
9.0	.1000D+01	.9744D+00	.1260D+00	.1143D–04	.1222D–12	.0000D+00
10.0	.1000D+01	.9966D+00	.4093D+00	.7858D–03	.5644D–09	.0000D+00
15.0	.1000D+01	.1000D+01	.9995D+00	.8204D+00	.6988D–01	.5482D–04
20.0	.1000D+01	.1000D+01	.1000D+01	.9999D+00	.9598D+00	.3732D+00
25.0	.1000D+01	.1000D+01	.1000D+01	.1000D+01	.1000D+01	.9922D+00
30.0	.1000D+01	.1000D+01	.1000D+01	.1000D+01	.1000D+01	.1000D+01
35.0	.1000D+01	.1000D+01	.1000D+01	.1000D+01	.1000D+01	.1000D+01
40.0	.1000D+01	.1000D+01	.1000D+01	.1000D+01	.1000D+01	.1000D+01
45.0	.1000D+01	.1000D+01	.1000D+01	.1000D+01	.1000D+01	.1000D+01
50.0	.1000D+01	.1000D+01	.1000D+01	.1000D+01	.1000D+01	.1000D+01
60.0	.1000D+01	.1000D+01	.1000D+01	.1000D+01	.1000D+01	.1000D+01

DIMENSIONLESS CONCENTRATION C/C0 FOR

V = .04 D = .04 R = 1.0 LAMBDA = .0 ALPHA = .0 T0 = 60.0

T(YEARS)	X = 350.	X = 400.	X = 450.	X = 500.	X = 550.	X = 600.
1.0	.0000D+00	.0000D+00	.0000D+00	.0000D+00	.0000D+00	.0000D+00
2.0	.0000D+00	.0000D+00	.0000D+00	.0000D+00	.0000D+00	.0000D+00
3.0	.0000D+00	.0000D+00	.0000D+00	.0000D+00	.0000D+00	.0000D+00
4.0	.0000D+00	.0000D+00	.0000D+00	.0000D+00	.0000D+00	.0000D+00
5.0	.0000D+00	.0000D+00	.0000D+00	.0000D+00	.0000D+00	.0000D+00
6.0	.0000D+00	.0000D+00	.0000D+00	.0000D+00	.0000D+00	.0000D+00
7.0	.0000D+00	.0000D+00	.0000D+00	.0000D+00	.0000D+00	.0000D+00
8.0	.0000D+00	.0000D+00	.0000D+00	.0000D+00	.0000D+00	.0000D+00
9.0	.0000D+00	.0000D+00	.0000D+00	.0000D+00	.0000D+00	.0000D+00
10.0	.0000D+00	.0000D+00	.0000D+00	.0000D+00	.0000D+00	.0000D+00
15.0	.1939D–09	.0000D+00	.0000D+00	.0000D+00	.0000D+00	.0000D+00
20.0	.8323D–02	.4006D–05	.3187D–10	.0000D+00	.0000D+00	.0000D+00
25.0	.7139D+00	.9898D–01	.8471D–03	.3010D–06	.3913D–11	.0000D+00
30.0	.9986D+00	.9024D+00	.3462D+00	.1849D–01	.7966D–04	.2302D–07
35.0	.1000D+01	.9998D+00	.9726D+00	.6388D+00	.1132D+00	.2765D–02
40.0	.1000D+01	.1000D+01	.1000D+01	.9933D+00	.8430D+00	.3239D+00
45.0	.1000D+01	.1000D+01	.1000D+01	.1000D+01	.9985D+00	.9436D+00
50.0	.1000D+01	.1000D+01	.1000D+01	.1000D+01	.1000D+01	.9997D+00
60.0	.1000D+01	.1000D+01	.1000D+01	.1000D+01	.1000D+01	.1000D+01

DIMENSIONLESS CONCENTRATION C/C0 FOR

V = .04 D = .40 R = 1.0 LAMBDA = .0 ALPHA = .0 T0 = 60.0

T(YEARS)	X = 50.	X = 100.	X = 150.	X = 200.	X = 250.	X = 300.
1.0	.1224D–01	.1258D–06	.0000D+00	.0000D+00	.0000D+00	.0000D+00
2.0	.1669D+00	.1135D–02	.1548D–06	.3516D–12	.0000D+00	.0000D+00
3.0	.3951D+00	.2324D–01	.1135D–03	.3826D–07	.8280D–12	.0000D+00
4.0	.5893D+00	.9949D–01	.2878D–02	.1180D–04	.6333D–08	.4293D–12
5.0	.7291D+00	.2257D+00	.1873D–01	.3413D–03	.1255D–05	.8955D–09
6.0	.8236D+00	.3726D+00	.6145D–01	.3004D–02	.3972D–04	.1357D–06
7.0	.8857D+00	.5140D+00	.1363D+00	.1337D–01	.4383D–03	.4578D–05
8.0	.9261D+00	.6357D+00	.2370D+00	.3881D–01	.2502D–02	.6022D–04
9.0	.9521D+00	.7336D+00	.3512D+00	.8483D–01	.9200D–02	.4222D–03
10.0	.9690D+00	.8086D+00	.4663D+00	.1522D+00	.2487D–01	.1906D–02
15.0	.9963D+00	.9684D+00	.8567D+00	.6138D+00	.3157D+00	.1067D+00
20.0	.9996D+00	.9953D+00	.9716D+00	.8901D+00	.7114D+00	.4572D+00
25.0	.9999D+00	.9993D+00	.9950D+00	.9756D+00	.9144D+00	.7798D+00
30.0	.1000D+01	.9999D+00	.9992D+00	.9953D+00	.9795D+00	.9327D+00
35.0	.1000D+01	.1000D+01	.9999D+00	.9991D+00	.9957D+00	.9829D+00
40.0	.1000D+01	.1000D+01	.1000D+01	.9999D+00	.9992D+00	.9962D+00
45.0	.1000D+01	.1000D+01	.1000D+01	.1000D+01	.9998D+00	.9992D+00
50.0	.1000D+01	.1000D+01	.1000D+01	.1000D+01	.1000D+01	.9998D+00
60.0	.1000D+01	.1000D+01	.1000D+01	.1000D+01	.1000D+01	.1000D+01

DIMENSIONLESS CONCENTRATION C/C0 FOR

V = .04 D = .40 R = 1.0 LAMBDA = .0 ALPHA = .0 T0 = 60.0

T(YEARS)	X = 350.	X = 400.	X = 450.	X = 500.	X = 550.	X = 600.
1.0	.0000D+00	.0000D+00	.0000D+00	.0000D+00	.0000D+00	.0000D+00
2.0	.0000D+00	.0000D+00	.0000D+00	.0000D+00	.0000D+00	.0000D+00
3.0	.0000D+00	.0000D+00	.0000D+00	.0000D+00	.0000D+00	.0000D+00
4.0	.0000D+00	.0000D+00	.0000D+00	.0000D+00	.0000D+00	.0000D+00
5.0	.1213D–12	.0000D+00	.0000D+00	.0000D+00	.0000D+00	.0000D+00
6.0	.1173D–09	.2528D–13	.0000D+00	.0000D+00	.0000D+00	.0000D+00
7.0	.1485D–07	.1474D–10	.0000D+00	.0000D+00	.0000D+00	.0000D+00
8.0	.5255D–06	.1638D–08	.1805D–11	.0000D+00	.0000D+00	.0000D+00
9.0	.7943D–05	.6021D–07	.1819D–09	.2175D–12	.0000D+00	.0000D+00
10.0	.6615D–04	.1019D–05	.6894D–08	.2031D–10	.2594D–13	.0000D+00
15.0	.2243D–01	.2840D–02	.2124D–03	.9272D–05	.2341D–06	.3408D–08
20.0	.2208D+00	.7643D–01	.1840D–01	.3020D–02	.3341D–03	.2471D–04
25.0	.5705D+00	.3394D+00	.1576D+00	.5551D–01	.1456D–01	.2806D–02
30.0	.8297D+00	.6596D+00	.4489D+00	.2523D+00	.1141D+00	.4072D–01
35.0	.9467D+00	.8669D+00	.7294D+00	.5444D+00	.3493D+00	.1880D+00
40.0	.9859D+00	.9576D+00	.8951D+00	.7842D+00	.6253D+00	.4416D+00
45.0	.9967D+00	.9884D+00	.9661D+00	.9169D+00	.8273D+00	.6927D+00
50.0	.9993D+00	.9971D+00	.9904D+00	.9728D+00	.9338D+00	.8615D+00
60.0	.1000D+01	.9999D+00	.9994D+00	.9979D+00	.9935D+00	.9823D+00

DIMENSIONLESS CONCENTRATION C/C0 FOR

V = .04 D = 2.00 R = 1.0 LAMBDA = .0 ALPHA = .0 T0 = 60.0

T(YEARS)	X = 50.	X = 100.	X = 150.	X = 200.	X = 250.	X = 300.
1.0	.9253D–01	.4885D–02	.5900D–04	.1482D–06	.7388D–10	.0000D+00
2.0	.2494D+00	.5540D–01	.6142D–02	.3210D–03	.7802D–05	.8437D–07
3.0	.3773D+00	.1396D+00	.3319D–01	.4872D–02	.4336D–03	.2288D–04
4.0	.4782D+00	.2310D+00	.8108D–01	.2007D–01	.3427D–02	.4021D–03
5.0	.5585D+00	.3181D+00	.1417D+00	.4813D–01	.1224D–01	.2314D–02
6.0	.6235D+00	.3973D+00	.2078D+00	.8729D–01	.2898D–01	.7510D–02
7.0	.6768D+00	.4678D+00	.2746D+00	.1344D+00	.5399D–01	.1762D–01
8.0	.7210D+00	.5298D+00	.3393D+00	.1862D+00	.8635D–01	.3351D–01
9.0	.7580D+00	.5842D+00	.4004D+00	.2402D+00	.1245D+00	.5529D–01
10.0	.7892D+00	.6319D+00	.4571D+00	.2944D+00	.1669D+00	.8250D–01
15.0	.8893D+00	.7958D+00	.6744D+00	.5356D+00	.3951D+00	.2688D+00
20.0	.9387D+00	.8832D+00	.8048D+00	.7049D+00	.5893D+00	.4672D+00
25.0	.9648D+00	.9315D+00	.8819D+00	.8143D+00	.7291D+00	.6298D+00
30.0	.9793D+00	.9591D+00	.9279D+00	.8832D+00	.8236D+00	.7492D+00
35.0	.9876D+00	.9752D+00	.9555D+00	.9263D+00	.8857D+00	.8325D+00
40.0	.9925D+00	.9848D+00	.9723D+00	.9533D+00	.9261D+00	.8889D+00
45.0	.9954D+00	.9906D+00	.9826D+00	.9703D+00	.9521D+00	.9266D+00
50.0	.9971D+00	.9941D+00	.9890D+00	.9810D+00	.9690D+00	.9516D+00
60.0	.9989D+00	.9976D+00	.9956D+00	.9922D+00	.9869D+00	.9790D+00

DIMENSIONLESS CONCENTRATION C/C0 FOR

V = .04 D = 2.00 R = 1.0 LAMBDA = .0 ALPHA = .0 T0 = 60.0

T(YEARS)	X = 350.	X = 400.	X = 450.	X = 500.	X = 550.	X = 600.
1.0	.0000D+00	.0000D+00	.0000D+00	.0000D+00	.0000D+00	.0000D+00
2.0	.4050D–09	.8535D–12	.0000D+00	.0000D+00	.0000D+00	.0000D+00
3.0	.7079D–06	.1277D–07	.1333D–09	.8031D–12	.0000D+00	.0000D+00
4.0	.3173D–04	.1677D–05	.5916D–07	.1387D–08	.2152D–10	.2208D–12
5.0	.3206D–03	.3239D–04	.2373D–05	.1258D–06	.4809D–08	.1323D–09
6.0	.1522D–02	.2369D–03	.2832D–04	.2586D–05	.1804D–06	.9581D–08
7.0	.4669D–02	.9896D–03	.1677D–03	.2264D–04	.2425D–05	.2061D–06
8.0	.1086D–01	.2903D–02	.6391D–03	.1156D–03	.1711D–04	.2069D–05
9.0	.2086D–01	.6711D–02	.1812D–02	.4112D–03	.7827D–04	.1247D–04
10.0	.3531D–01	.1311D–01	.4169D–02	.1135D–02	.2640D–03	.5238D–04
15.0	.1676D+00	.9525D–01	.4950D–01	.2324D–01	.9865D–02	.3780D–02
20.0	.3495D+00	.2455D+00	.1619D+00	.9949D–01	.5690D–01	.3024D–01
25.0	.5224D+00	.4143D+00	.3140D+00	.2257D+00	.1538D+00	.9915D–01
30.0	.6622D+00	.5666D+00	.4687D+00	.3726D+00	.2842D+00	.2077D+00
35.0	.7665D+00	.6892D+00	.6043D+00	.5140D+00	.4234D+00	.3371D+00
40.0	.8409D+00	.7816D+00	.7132D+00	.6357D+00	.5528D+00	.4679D+00
45.0	.8925D+00	.8487D+00	.7962D+00	.7336D+00	.6630D+00	.5866D+00
50.0	.9278D+00	.8962D+00	.8571D+00	.8086D+00	.7515D+00	.6869D+00
60.0	.9677D+00	.9524D+00	.9316D+00	.9048D+00	.8711D+00	.8302D+00

DIMENSIONLESS CONCENTRATION C/C0 FOR

V = .06 D = .06 R = 1.0 LAMBDA = .0 ALPHA = .0 T0 = 60.0

T(YEARS)	X = 50.	X = 100.	X = 150.	X = 200.	X = 250.	X = 300.
1.0	.9272D–05	.0000D+00	.0000D+00	.0000D+00	.0000D+00	.0000D+00
2.0	.2523D+00	.8342D–09	.0000D+00	.0000D+00	.0000D+00	.0000D+00
3.0	.9169D+00	.1340D–02	.8510D–13	.0000D+00	.0000D+00	.0000D+00
4.0	.9979D+00	.1743D+00	.1159D–05	.0000D+00	.0000D+00	.0000D+00
5.0	.1000D+01	.7419D+00	.3072D–02	.4592D–09	.0000D+00	.0000D+00
6.0	.1000D+01	.9744D+00	.1260D+00	.1143D–04	.1222D–12	.0000D+00
7.0	.1000D+01	.9989D+00	.5772D+00	.3827D–02	.1643D–07	.0000D+00
8.0	.1000D+01	.1000D+01	.9125D+00	.9317D–01	.3221D–04	.1287D–10
9.0	.1000D+01	.1000D+01	.9915D+00	.4444D+00	.3883D–02	.1094D–06
10.0	.1000D+01	.1000D+01	.9995D+00	.8204D+00	.6988D–01	.5482D–04
15.0	.1000D+01	.1000D+01	.1000D+01	.1000D+01	.9990D+00	.8691D+00
20.0	.1000D+01	.1000D+01	.1000D+01	.1000D+01	.1000D+01	.1000D+01
25.0	.1000D+01	.1000D+01	.1000D+01	.1000D+01	.1000D+01	.1000D+01
30.0	.1000D+01	.1000D+01	.1000D+01	.1000D+01	.1000D+01	.1000D+01
35.0	.1000D+01	.1000D+01	.1000D+01	.1000D+01	.1000D+01	.1000D+01
40.0	.1000D+01	.1000D+01	.1000D+01	.1000D+01	.1000D+01	.1000D+01
45.0	.1000D+01	.1000D+01	.1000D+01	.1000D+01	.1000D+01	.1000D+01
50.0	.1000D+01	.1000D+01	.1000D+01	.1000D+01	.1000D+01	.1000D+01
60.0	.1000D+01	.1000D+01	.1000D+01	.1000D+01	.1000D+01	.1000D+01

DIMENSIONLESS CONCENTRATION C/C0 FOR

V = .06 D = .06 R = 1.0 LAMBDA = .0 ALPHA = .0 T0 = 60.0

T(YEARS)	X = 350.	X = 400.	X = 450.	X = 500.	X = 550.	X = 600.
1.0	.0000D+00	.0000D+00	.0000D+00	.0000D+00	.0000D+00	.0000D+00
2.0	.0000D+00	.0000D+00	.0000D+00	.0000D+00	.0000D+00	.0000D+00
3.0	.0000D+00	.0000D+00	.0000D+00	.0000D+00	.0000D+00	.0000D+00
4.0	.0000D+00	.0000D+00	.0000D+00	.0000D+00	.0000D+00	.0000D+00
5.0	.0000D+00	.0000D+00	.0000D+00	.0000D+00	.0000D+00	.0000D+00
6.0	.0000D+00	.0000D+00	.0000D+00	.0000D+00	.0000D+00	.0000D+00
7.0	.0000D+00	.0000D+00	.0000D+00	.0000D+00	.0000D+00	.0000D+00
8.0	.0000D+00	.0000D+00	.0000D+00	.0000D+00	.0000D+00	.0000D+00
9.0	.0000D+00	.0000D+00	.0000D+00	.0000D+00	.0000D+00	.0000D+00
10.0	.1939D–09	.0000D+00	.0000D+00	.0000D+00	.0000D+00	.0000D+00
15.0	.2030D+00	.2691D–02	.1095D–05	.1144D–10	.0000D+00	.0000D+00
20.0	.9986D+00	.9024D+00	.3462D+00	.1849D–01	.7966D–04	.2302D–07
25.0	.1000D+01	.1000D+01	.9985D+00	.9261D+00	.4744D+00	.5751D–01
30.0	.1000D+01	.1000D+01	.1000D+01	.1000D+01	.9985D+00	.9436D+00
35.0	.1000D+01	.1000D+01	.1000D+01	.1000D+01	.1000D+01	.1000D+01
40.0	.1000D+01	.1000D+01	.1000D+01	.1000D+01	.1000D+01	.1000D+01
45.0	.1000D+01	.1000D+01	.1000D+01	.1000D+01	.1000D+01	.1000D+01
50.0	.1000D+01	.1000D+01	.1000D+01	.1000D+01	.1000D+01	.1000D+01
60.0	.1000D+01	.1000D+01	.1000D+01	.1000D+01	.1000D+01	.1000D+01

DIMENSIONLESS CONCENTRATION C/C0 FOR

V = .06 D = .60 R = 1.0 LAMBDA = .0 ALPHA = .0 T0 = 60.0

T(YEARS)	X = 50.	X = 100.	X = 150.	X = 200.	X = 250.	X = 300.
1.0	.6934D–01	.5398D–04	.2054D–09	.0000D+00	.0000D+00	.0000D+00
2.0	.3951D+00	.2324D–01	.1135D–03	.3826D–07	.8280D–12	.0000D+00
3.0	.6657D+00	.1580D+00	.8227D–02	.7735D–04	.1210D–06	.3033D–10
4.0	.8236D+00	.3726D+00	.6145D–01	.3004D–02	.3972D–04	.1357D–06
5.0	.9081D+00	.5778D+00	.1841D+00	.2377D–01	.1119D–02	.1825D–04
6.0	.9521D+00	.7336D+00	.3512D+00	.8483D–01	.9200D–02	.4222D–03
7.0	.9750D+00	.8386D+00	.5212D+00	.1928D+00	.3746D–01	.3570D–02
8.0	.9869D+00	.9048D+00	.6666D+00	.3324D+00	.9858D–01	.1611D–01
9.0	.9931D+00	.9448D+00	.7778D+00	.4793D+00	.1946D+00	.4790D–01
10.0	.9963D+00	.9684D+00	.8567D+00	.6138D+00	.3157D+00	.1067D+00
15.0	.9998D+00	.9982D+00	.9880D+00	.9470D+00	.8375D+00	.6391D+00
20.0	.1000D+01	.9999D+00	.9992D+00	.9953D+00	.9795D+00	.9327D+00
25.0	.1000D+01	.1000D+01	.9999D+00	.9996D+00	.9981D+00	.9918D+00
30.0	.1000D+01	.1000D+01	.1000D+01	.1000D+01	.9998D+00	.9992D+00
35.0	.1000D+01	.1000D+01	.1000D+01	.1000D+01	.1000D+01	.9999D+00
40.0	.1000D+01	.1000D+01	.1000D+01	.1000D+01	.1000D+01	.1000D+01
45.0	.1000D+01	.1000D+01	.1000D+01	.1000D+01	.1000D+01	.1000D+01
50.0	.1000D+01	.1000D+01	.1000D+01	.1000D+01	.1000D+01	.1000D+01
60.0	.1000D+01	.1000D+01	.1000D+01	.1000D+01	.1000D+01	.1000D+01

DIMENSIONLESS CONCENTRATION C/C0 FOR

V = .06 D = .60 R = 1.0 LAMBDA = .0 ALPHA = .0 T0 = 60.0

T(YEARS)	X = 350.	X = 400.	X = 450.	X = 500.	X = 550.	X = 600.
1.0	.0000D+00	.0000D+00	.0000D+00	.0000D+00	.0000D+00	.0000D+00
2.0	.0000D+00	.0000D+00	.0000D+00	.0000D+00	.0000D+00	.0000D+00
3.0	.0000D+00	.0000D+00	.0000D+00	.0000D+00	.0000D+00	.0000D+00
4.0	.1173D–09	.2528D–13	.0000D+00	.0000D+00	.0000D+00	.0000D+00
5.0	.1004D–06	.1834D–09	.1103D–12	.0000D+00	.0000D+00	.0000D+00
6.0	.7943D–05	.6021D–07	.1819D–09	.2175D–12	.0000D+00	.0000D+00
7.0	.1609D–03	.3359D–05	.3208D–07	.1391D–09	.2723D–12	.0000D+00
8.0	.1390D–02	.6181D–04	.1396D–05	.1588D–07	.9038D–10	.2565D–12
9.0	.6805D–02	.5425D–03	.2388D–04	.5737D–06	.7481D–08	.5269D–10
10.0	.2243D–01	.2840D–02	.2124D–03	.9272D–05	.2341D–06	.3408D–08
15.0	.3939D+00	.1861D+00	.6504D–01	.1641D–01	.2944D–02	.3717D–03
20.0	.8297D+00	.6596D+00	.4489D+00	.2523D+00	.1141D+00	.4072D–01
25.0	.9721D+00	.9232D+00	.8271D+00	.6772D+00	.4912D+00	.3073D+00
30.0	.9967D+00	.9884D+00	.9661D+00	.9169D+00	.8273D+00	.6927D+00
35.0	.9997D+00	.9986D+00	.9951D+00	.9852D+00	.9613D+00	.9128D+00
40.0	.1000D+01	.9999D+00	.9994D+00	.9979D+00	.9935D+00	.9823D+00
45.0	.1000D+01	.1000D+01	.9999D+00	.9998D+00	.9991D+00	.9972D+00
50.0	.1000D+01	.1000D+01	.1000D+01	.1000D+01	.9999D+00	.9996D+00
60.0	.1000D+01	.1000D+01	.1000D+01	.1000D+01	.1000D+01	.1000D+01

DIMENSIONLESS CONCENTRATION C/C0 FOR

V = .06 D = 3.00 R = 1.0 LAMBDA = .0 ALPHA = .0 T0 = 60.0

T(YEARS)	X = 50.	X = 100.	X = 150.	X = 200.	X = 250.	X = 300.
1.0	.1737D+00	.2363D–01	.1239D–02	.2360D–04	.1555D–06	.3477D–09
2.0	.3773D+00	.1396D+00	.3319D–01	.4872D–02	.4336D–03	.2288D–04
3.0	.5205D+00	.2754D+00	.1103D+00	.3254D–01	.6931D–02	.1060D–02
4.0	.6235D+00	.3973D+00	.2078D+00	.8729D–01	.2898D–01	.7510D–02
5.0	.6999D+00	.4998D+00	.3074D+00	.1599D+00	.6934D–01	.2482D–01
6.0	.7580D+00	.5842D+00	.4004D+00	.2402D+00	.1245D+00	.5529D–01
7.0	.8030D+00	.6535D+00	.4838D+00	.3212D+00	.1891D+00	.9790D–01
8.0	.8384D+00	.7104D+00	.5568D+00	.3987D+00	.2581D+00	.1498D+00
9.0	.8666D+00	.7572D+00	.6200D+00	.4705D+00	.3277D+00	.2078D+00
10.0	.8893D+00	.7958D+00	.6744D+00	.5356D+00	.3951D+00	.2688D+00
15.0	.9537D+00	.9108D+00	.8484D+00	.7658D+00	.6657D+00	.5540D+00
20.0	.9793D+00	.9591D+00	.9279D+00	.8832D+00	.8236D+00	.7492D+00
25.0	.9904D+00	.9806D+00	.9649D+00	.9414D+00	.9081D+00	.8635D+00
30.0	.9954D+00	.9906D+00	.9826D+00	.9703D+00	.9521D+00	.9266D+00
35.0	.9977D+00	.9953D+00	.9913D+00	.9848D+00	.9750D+00	.9607D+00
40.0	.9989D+00	.9976D+00	.9956D+00	.9922D+00	.9869D+00	.9790D+00
45.0	.9994D+00	.9988D+00	.9977D+00	.9959D+00	.9931D+00	.9887D+00
50.0	.9997D+00	.9994D+00	.9988D+00	.9979D+00	.9963D+00	.9939D+00
60.0	.9999D+00	.9998D+00	.9997D+00	.9994D+00	.9990D+00	.9983D+00

DIMENSIONLESS CONCENTRATION C/C0 FOR

V = .06 D = 3.00 R = 1.0 LAMBDA = .0 ALPHA = .0 T0 = 60.0

T(YEARS)	X = 350.	X = 400.	X = 450.	X = 500.	X = 550.	X = 600.
1.0	.2598D–12	.0000D+00	.0000D+00	.0000D+00	.0000D+00	.0000D+00
2.0	.7079D–06	.1277D–07	.1333D–09	.8031D–12	.0000D+00	.0000D+00
3.0	.1143D–03	.8665D–05	.4583D–06	.1690D–07	.4328D–09	.7678D–11
4.0	.1522D–02	.2369D–03	.2832D–04	.2586D–05	.1804D–06	.9581D–08
5.0	.7321D–02	.1756D–02	.3422D–03	.5398D–04	.6878D–05	.7049D–06
6.0	.2086D–01	.6711D–02	.1812D–02	.4112D–03	.7827D–04	.1247D–04
7.0	.4423D–01	.1747D–01	.5955D–02	.1753D–02	.4443D–03	.9685D–04
8.0	.7746D–01	.3568D–01	.1448D–01	.5179D–02	.1628D–02	.4490D–03
9.0	.1192D+00	.6187D–01	.2876D–01	.1196D–01	.4445D–02	.1472D–02
10.0	.1676D+00	.9525D–01	.4950D–01	.2324D–01	.9865D–02	.3780D–02
15.0	.4393D+00	.3304D+00	.2356D+00	.1580D+00	.9968D–01	.5900D–01
20.0	.6622D+00	.5666D+00	.4687D+00	.3726D+00	.2842D+00	.2077D+00
25.0	.8070D+00	.7389D+00	.6622D+00	.5778D+00	.4901D+00	.4033D+00
30.0	.8925D+00	.8487D+00	.7962D+00	.7336D+00	.6630D+00	.5866D+00
35.0	.9409D+00	.9142D+00	.8808D+00	.8386D+00	.7881D+00	.7297D+00
40.0	.9677D+00	.9524D+00	.9316D+00	.9048D+00	.8711D+00	.8302D+00
45.0	.9826D+00	.9736D+00	.9613D+00	.9448D+00	.9234D+00	.8964D+00
50.0	.9905D+00	.9854D+00	.9783D+00	.9684D+00	.9552D+00	.9380D+00
60.0	.9972D+00	.9956D+00	.9932D+00	.9899D+00	.9851D+00	.9787D+00

DIMENSIONLESS CONCENTRATION C/C0 FOR

V = .08 D = .08 R = 1.0 LAMBDA = .0 ALPHA = .0 T0 = 60.0

T(YEARS)	X = 50.	X = 100.	X = 150.	X = 200.	X = 250.	X = 300.
1.0	.3020D–02	.0000D+00	.0000D+00	.0000D+00	.0000D+00	.0000D+00
2.0	.7842D+00	.5599D–04	.0000D+00	.0000D+00	.0000D+00	.0000D+00
3.0	.9979D+00	.1743D+00	.1159D–05	.0000D+00	.0000D+00	.0000D+00
4.0	.1000D+01	.8662D+00	.1486D–01	.2522D–07	.0000D+00	.0000D+00
5.0	.1000D+01	.9966D+00	.4093D+00	.7858D–03	.5644D–09	.0000D+00
6.0	.1000D+01	.1000D+01	.9125D+00	.9317D–01	.3221D–04	.1287D–10
7.0	.1000D+01	.1000D+01	.9966D+00	.5889D+00	.1215D–01	.1136D–05
8.0	.1000D+01	.1000D+01	.9999D+00	.9412D+00	.2258D+00	.1075D–02
9.0	.1000D+01	.1000D+01	.1000D+01	.9970D+00	.7146D+00	.5294D–01
10.0	.1000D+01	.1000D+01	.1000D+01	.9999D+00	.9598D+00	.3732D+00
15.0	.1000D+01	.1000D+01	.1000D+01	.1000D+01	.1000D+01	.1000D+01
20.0	.1000D+01	.1000D+01	.1000D+01	.1000D+01	.1000D+01	.1000D+01
25.0	.1000D+01	.1000D+01	.1000D+01	.1000D+01	.1000D+01	.1000D+01
30.0	.1000D+01	.1000D+01	.1000D+01	.1000D+01	.1000D+01	.1000D+01
35.0	.1000D+01	.1000D+01	.1000D+01	.1000D+01	.1000D+01	.1000D+01
40.0	.1000D+01	.1000D+01	.1000D+01	.1000D+01	.1000D+01	.1000D+01
45.0	.1000D+01	.1000D+01	.1000D+01	.1000D+01	.1000D+01	.1000D+01
50.0	.1000D+01	.1000D+01	.1000D+01	.1000D+01	.1000D+01	.1000D+01
60.0	.1000D+01	.1000D+01	.1000D+01	.1000D+01	.1000D+01	.1000D+01

DIMENSIONLESS CONCENTRATION C/C0 FOR

V = .08 D = .08 R = 1.0 LAMBDA = .0 ALPHA = .0 T0 = 60.0

T(YEARS)	X = 350.	X = 400.	X = 450.	X = 500.	X = 550.	X = 600.
1.0	.0000D+00	.0000D+00	.0000D+00	.0000D+00	.0000D+00	.0000D+00
2.0	.0000D+00	.0000D+00	.0000D+00	.0000D+00	.0000D+00	.0000D+00
3.0	.0000D+00	.0000D+00	.0000D+00	.0000D+00	.0000D+00	.0000D+00
4.0	.0000D+00	.0000D+00	.0000D+00	.0000D+00	.0000D+00	.0000D+00
5.0	.0000D+00	.0000D+00	.0000D+00	.0000D+00	.0000D+00	.0000D+00
6.0	.0000D+00	.0000D+00	.0000D+00	.0000D+00	.0000D+00	.0000D+00
7.0	.2971D–12	.0000D+00	.0000D+00	.0000D+00	.0000D+00	.0000D+00
8.0	.3652D–07	.0000D+00	.0000D+00	.0000D+00	.0000D+00	.0000D+00
9.0	.7242D–04	.1102D–08	.0000D+00	.0000D+00	.0000D+00	.0000D+00
10.0	.8323D–02	.4006D–05	.3187D–10	.0000D+00	.0000D+00	.0000D+00
15.0	.9986D+00	.9024D+00	.3462D+00	.1849D–01	.7966D–04	.2302D–07
20.0	.1000D+01	.1000D+01	.1000D+01	.9933D+00	.8430D+00	.3239D+00
25.0	.1000D+01	.1000D+01	.1000D+01	.1000D+01	.1000D+01	.9997D+00
30.0	.1000D+01	.1000D+01	.1000D+01	.1000D+01	.1000D+01	.1000D+01
35.0	.1000D+01	.1000D+01	.1000D+01	.1000D+01	.1000D+01	.1000D+01
40.0	.1000D+01	.1000D+01	.1000D+01	.1000D+01	.1000D+01	.1000D+01
45.0	.1000D+01	.1000D+01	.1000D+01	.1000D+01	.1000D+01	.1000D+01
50.0	.1000D+01	.1000D+01	.1000D+01	.1000D+01	.1000D+01	.1000D+01
60.0	.1000D+01	.1000D+01	.1000D+01	.1000D+01	.1000D+01	.1000D+01

DIMENSIONLESS CONCENTRATION C/C0 FOR

V = .08 D = .80 R = 1.0 LAMBDA = .0 ALPHA = .0 T0 = 60.0

T(YEARS)	X = 50.	X = 100.	X = 150.	X = 200.	X = 250.	X = 300.
1.0	.1669D+00	.1135D–02	.1548D–06	.3516D–12	.0000D+00	.0000D+00
2.0	.5893D+00	.9949D–01	.2878D–02	.1180D–04	.6333D–08	.4293D–12
3.0	.8236D+00	.3726D+00	.6145D–01	.3004D–02	.3972D–04	.1357D–06
4.0	.9261D+00	.6357D+00	.2370D+00	.3881D–01	.2502D–02	.6022D–04
5.0	.9690D+00	.8086D+00	.4663D+00	.1522D+00	.2487D–01	.1906D–02
6.0	.9869D+00	.9048D+00	.6666D+00	.3324D+00	.9858D–01	.1611D–01
7.0	.9944D+00	.9541D+00	.8074D+00	.5262D+00	.2329D+00	.6429D–01
8.0	.9976D+00	.9783D+00	.8946D+00	.6909D+00	.4024D+00	.1616D+00
9.0	.9990D+00	.9899D+00	.9445D+00	.8110D+00	.5704D+00	.3007D+00
10.0	.9996D+00	.9953D+00	.9716D+00	.8901D+00	.7114D+00	.4572D+00
15.0	.1000D+01	.9999D+00	.9992D+00	.9953D+00	.9795D+00	.9327D+00
20.0	.1000D+01	.1000D+01	.1000D+01	.9999D+00	.9992D+00	.9962D+00
25.0	.1000D+01	.1000D+01	.1000D+01	.1000D+01	.1000D+01	.9998D+00
30.0	.1000D+01	.1000D+01	.1000D+01	.1000D+01	.1000D+01	.1000D+01
35.0	.1000D+01	.1000D+01	.1000D+01	.1000D+01	.1000D+01	.1000D+01
40.0	.1000D+01	.1000D+01	.1000D+01	.1000D+01	.1000D+01	.1000D+01
45.0	.1000D+01	.1000D+01	.1000D+01	.1000D+01	.1000D+01	.1000D+01
50.0	.1000D+01	.1000D+01	.1000D+01	.1000D+01	.1000D+01	.1000D+01
60.0	.1000D+01	.1000D+01	.1000D+01	.1000D+01	.1000D+01	.1000D+01

DIMENSIONLESS CONCENTRATION C/C0 FOR

V = .08 D = .80 R = 1.0 LAMBDA = .0 ALPHA = .0 T0 = 60.0

T(YEARS)	X = 350.	X = 400.	X = 450.	X = 500.	X = 550.	X = 600.
1.0	.0000D+00	.0000D+00	.0000D+00	.0000D+00	.0000D+00	.0000D+00
2.0	.0000D+00	.0000D+00	.0000D+00	.0000D+00	.0000D+00	.0000D+00
3.0	.1173D–09	.2528D–13	.0000D+00	.0000D+00	.0000D+00	.0000D+00
4.0	.5255D–06	.1638D–08	.1805D–11	.0000D+00	.0000D+00	.0000D+00
5.0	.6615D–04	.1019D–05	.6894D–08	.2031D–10	.2594D–13	.0000D+00
6.0	.1390D–02	.6181D–04	.1396D–05	.1588D–07	.9038D–10	.2565D–12
7.0	.1051D–01	.9894D–03	.5265D–04	.1566D–05	.2586D–07	.2361D–09
8.0	.4217D–01	.6906D–02	.6945D–03	.4234D–04	.1550D–05	.3392D–07
9.0	.1114D+00	.2780D–01	.4558D–02	.4831D–03	.3278D–04	.1412D–05
10.0	.2208D+00	.7643D–01	.1840D–01	.3020D–02	.3341D–03	.2471D–04
15.0	.8297D+00	.6596D+00	.4489D+00	.2523D+00	.1141D+00	.4072D–01
20.0	.9859D+00	.9576D+00	.8951D+00	.7842D+00	.6253D+00	.4416D+00
25.0	.9993D+00	.9971D+00	.9904D+00	.9728D+00	.9338D+00	.8615D+00
30.0	.1000D+01	.9999D+00	.9994D+00	.9979D+00	.9935D+00	.9823D+00
35.0	.1000D+01	.1000D+01	.1000D+01	.9999D+00	.9996D+00	.9985D+00
40.0	.1000D+01	.1000D+01	.1000D+01	.1000D+01	.1000D+01	.9999D+00
45.0	.1000D+01	.1000D+01	.1000D+01	.1000D+01	.1000D+01	.1000D+01
50.0	.1000D+01	.1000D+01	.1000D+01	.1000D+01	.1000D+01	.1000D+01
60.0	.1000D+01	.1000D+01	.1000D+01	.1000D+01	.1000D+01	.1000D+01

DIMENSIONLESS CONCENTRATION C/C0 FOR

V = .08 D = 4.00 R = 1.0 LAMBDA = .0 ALPHA = .0 T0 = 60.0

T(YEARS)	X = 50.	X = 100.	X = 150.	X = 200.	X = 250.	X = 300.
1.0	.2494D+00	.5540D-01	.6142D-02	.3210D-03	.7802D-05	.8437D-07
2.0	.4782D+00	.2310D+00	.8108D-01	.2007D-01	.3427D-02	.4021D-03
3.0	.6235D+00	.3973D+00	.2078D+00	.8729D-01	.2898D-01	.7510D-02
4.0	.7210D+00	.5298D+00	.3393D+00	.1862D+00	.8635D-01	.3351D-01
5.0	.7892D+00	.6319D+00	.4571D+00	.2944D+00	.1669D+00	.8250D-01
6.0	.8384D+00	.7104D+00	.5568D+00	.3987D+00	.2581D+00	.1498D+00
7.0	.8747D+00	.7709D+00	.6391D+00	.4929D+00	.3505D+00	.2279D+00
8.0	.9020D+00	.8178D+00	.7062D+00	.5752D+00	.4381D+00	.3099D+00
9.0	.9227D+00	.8544D+00	.7607D+00	.6455D+00	.5181D+00	.3908D+00
10.0	.9387D+00	.8832D+00	.8048D+00	.7049D+00	.5893D+00	.4672D+00
15.0	.9793D+00	.9591D+00	.9279D+00	.8832D+00	.8236D+00	.7492D+00
20.0	.9925D+00	.9848D+00	.9723D+00	.9533D+00	.9261D+00	.8889D+00
25.0	.9971D+00	.9941D+00	.9890D+00	.9810D+00	.9690D+00	.9516D+00
30.0	.9989D+00	.9976D+00	.9956D+00	.9922D+00	.9869D+00	.9790D+00
35.0	.9995D+00	.9990D+00	.9982D+00	.9967D+00	.9944D+00	.9908D+00
40.0	.9998D+00	.9996D+00	.9992D+00	.9986D+00	.9976D+00	.9961D+00
45.0	.9999D+00	.9998D+00	.9997D+00	.9994D+00	.9990D+00	.9983D+00
50.0	.1000D+01	.9999D+00	.9999D+00	.9997D+00	.9996D+00	.9992D+00
60.0	.1000D+01	.1000D+01	.1000D+01	.1000D+01	.9999D+00	.9999D+00

DIMENSIONLESS CONCENTRATION C/C0 FOR

V = .08 D = 4.00 R = 1.0 LAMBDA = .0 ALPHA = .0 T0 = 60.0

T(YEARS)	X = 350.	X = 400.	X = 450.	X = 500.	X = 550.	X = 600.
1.0	.4050D-09	.8535D-12	.0000D+00	.0000D+00	.0000D+00	.0000D+00
2.0	.3173D-04	.1677D-05	.5916D-07	.1387D-08	.2152D-10	.2208D-12
3.0	.1522D-02	.2369D-03	.2832D-04	.2586D-05	.1804D-06	.9581D-08
4.0	.1086D-01	.2903D-02	.6391D-03	.1156D-03	.1711D-04	.2069D-05
5.0	.3531D-01	.1311D-01	.4169D-02	.1135D-02	.2640D-03	.5238D-04
6.0	.7746D-01	.3568D-01	.1448D-01	.5179D-02	.1628D-02	.4490D-03
7.0	.1347D+00	.7202D-01	.3494D-01	.1518D-01	.5914D-02	.2063D-02
8.0	.2023D+00	.1213D+00	.6697D-01	.3364D-01	.1539D-01	.6402D-02
9.0	.2753D+00	.1804D+00	.1100D+00	.6180D-01	.3202D-01	.1527D-01
10.0	.3495D+00	.2455D+00	.1619D+00	.9949D-01	.5690D-01	.3024D-01
15.0	.6622D+00	.5666D+00	.4687D+00	.3726D+00	.2842D+00	.2077D+00
20.0	.8409D+00	.7816D+00	.7132D+00	.6357D+00	.5528D+00	.4679D+00
25.0	.9278D+00	.8962D+00	.8571D+00	.8086D+00	.7515D+00	.6869D+00
30.0	.9677D+00	.9524D+00	.9316D+00	.9048D+00	.8711D+00	.8302D+00
35.0	.9858D+00	.9784D+00	.9680D+00	.9541D+00	.9358D+00	.9125D+00
40.0	.9937D+00	.9902D+00	.9852D+00	.9783D+00	.9689D+00	.9564D+00
45.0	.9972D+00	.9956D+00	.9932D+00	.9899D+00	.9851D+00	.9787D+00
50.0	.9988D+00	.9980D+00	.9969D+00	.9953D+00	.9930D+00	.9897D+00
60.0	.9998D+00	.9996D+00	.9994D+00	.9990D+00	.9985D+00	.9977D+00

DIMENSIONLESS CONCENTRATION C/C0 FOR

V = .10 D = .10 R = 1.0 LAMBDA = .0 ALPHA = .0 T0 = 60.0

T(YEARS)	X = 50.	X = 100.	X = 150.	X = 200.	X = 250.	X = 300.
.5	.5843D–0	.0000D+00	.0000D+00	.0000D+00	.0000D+00	.0000D+00
1.0	.5551D–01	.4338D–13	.0000D+00	.0000D+00	.0000D+00	.0000D+00
2.0	.9728D+00	.1251D–01	.8496D–10	.0000D+00	.0000D+00	.0000D+00
3.0	.1000D+01	.7419D+00	.3072D–02	.4592D–09	.0000D+00	.0000D+00
4.0	.1000D+01	.9966D+00	.4093D+00	.7858D–03	.5644D–09	.0000D+00
5.0	.1000D+01	.1000D+01	.9566D+00	.1810D+00	.2059D–03	.3835D–09
6.0	.1000D+01	.1000D+01	.9995D+00	.8204D+00	.6988D–01	.5482D–04
7.0	.1000D+01	.1000D+01	.1000D+01	.9932D+00	.5992D+00	.2479D–01
8.0	.1000D+01	.1000D+01	.1000D+01	.9999D+00	.9598D+00	.3732D+00
9.0	.1000D+01	.1000D+01	.1000D+01	.1000D+01	.9990D+00	.8691D+00
10.0	.1000D+01	.1000D+01	.1000D+01	.1000D+01	.1000D+01	.9922D+00
15.0	.1000D+01	.1000D+01	.1000D+01	.1000D+01	.1000D+01	.1000D+01
20.0	.1000D+01	.1000D+01	.1000D+01	.1000D+01	.1000D+01	.1000D+01
25.0	.1000D+01	.1000D+01	.1000D+01	.1000D+01	.1000D+01	.1000D+01
30.0	.1000D+01	.1000D+01	.1000D+01	.1000D+01	.1000D+01	.1000D+01
35.0	.1000D+01	.1000D+01	.1000D+01	.1000D+01	.1000D+01	.1000D+01
40.0	.1000D+01	.1000D+01	.1000D+01	.1000D+01	.1000D+01	.1000D+01
45.0	.1000D+01	.1000D+01	.1000D+01	.1000D+01	.1000D+01	.1000D+01
50.0	.1000D+01	.1000D+01	.1000D+01	.1000D+01	.1000D+01	.1000D+01

DIMENSIONLESS CONCENTRATION C/C0 FOR

V = .10 D = .10 R = 1.0 LAMBDA = .0 ALPHA = .0 T0 = 60.0

T(YEARS)	X = 400.	X = 500.	X = 600.	X = 700.	X = 800.	X = 1000.
.5	.0000D+00	.0000D+00	.0000D+00	.0000D+00	.0000D+00	.0000D+00
1.0	.0000D+00	.0000D+00	.0000D+00	.0000D+00	.0000D+00	.0000D+00
2.0	.0000D+00	.0000D+00	.0000D+00	.0000D+00	.0000D+00	.0000D+00
3.0	.0000D+00	.0000D+00	.0000D+00	.0000D+00	.0000D+00	.0000D+00
4.0	.0000D+00	.0000D+00	.0000D+00	.0000D+00	.0000D+00	.0000D+00
5.0	.0000D+00	.0000D+00	.0000D+00	.0000D+00	.0000D+00	.0000D+00
6.0	.0000D+00	.0000D+00	.0000D+00	.0000D+00	.0000D+00	.0000D+00
7.0	.8283D–10	.0000D+00	.0000D+00	.0000D+00	.0000D+00	.0000D+00
8.0	.4006D–05	.0000D+00	.0000D+00	.0000D+00	.0000D+00	.0000D+00
9.0	.2691D–02	.1144D–10	.0000D+00	.0000D+00	.0000D+00	.0000D+00
10.0	.9898D–01	.3010D–06	.0000D+00	.0000D+00	.0000D+00	.0000D+00
15.0	.1000D+01	.9261D+00	.5751D–01	.2123D–05	.1256D–13	.0000D+00
20.0	.1000D+01	.1000D+01	.9997D+00	.7877D+00	.3442D–01	.8679D–12
25.0	.1000D+01	.1000D+01	.1000D+01	.1000D+01	.9960D+00	.2098D–01
30.0	.1000D+01	.1000D+01	.1000D+01	.1000D+01	.1000D+01	.9796D+00
35.0	.1000D+01	.1000D+01	.1000D+01	.1000D+01	.1000D+01	.1000D+01
40.0	.1000D+01	.1000D+01	.1000D+01	.1000D+01	.1000D+01	.1000D+01
45.0	.1000D+01	.1000D+01	.1000D+01	.1000D+01	.1000D+01	.1000D+01
50.0	.1000D+01	.1000D+01	.1000D+01	.1000D+01	.1000D+01	.1000D+01

DIMENSIONLESS CONCENTRATION C/C0 FOR

V = .10 D = 1.00 R = 1.0 LAMBDA = .0 ALPHA = .0 T0 = 60.0

T(YEARS)	X = 50.	X = 100.	X = 150.	X = 200.	X = 250.	X = 300.
.5	.3448D–01	.4743D–05	.1032D–11	.0000D+00	.0000D+00	.0000D+00
1.0	.2814D+00	.7006D–02	.8193D–05	.3736D–09	.0000D+00	.0000D+00
2.0	.7291D+00	.2257D+00	.1873D–01	.3413D–03	.1255D–05	.8955D–09
3.0	.9081D+00	.5778D+00	.1841D+00	.2377D–01	.1119D–02	.1825D–04
4.0	.9690D+00	.8086D+00	.4663D+00	.1522D+00	.2487D–01	.1906D–02
5.0	.9894D+00	.9205D+00	.7075D+00	.3817D+00	.1270D+00	.2408D–01
6.0	.9963D+00	.9684D+00	.8567D+00	.6138D+00	.3157D+00	.1067D+00
7.0	.9987D+00	.9877D+00	.9347D+00	.7852D+00	.5302D+00	.2632D+00
8.0	.9996D+00	.9953D+00	.9716D+00	.8901D+00	.7114D+00	.4572D+00
9.0	.9998D+00	.9982D+00	.9880D+00	.9470D+00	.8375D+00	.6391D+00
10.0	.9999D+00	.9993D+00	.9950D+00	.9756D+00	.9144D+00	.7798D+00
15.0	.1000D+01	.1000D+01	.9999D+00	.9996D+00	.9981D+00	.9918D+00
20.0	.1000D+01	.1000D+01	.1000D+01	.1000D+01	.1000D+01	.9998D+00
25.0	.1000D+01	.1000D+01	.1000D+01	.1000D+01	.1000D+01	.1000D+01
30.0	.1000D+01	.1000D+01	.1000D+01	.1000D+01	.1000D+01	.1000D+01
35.0	.1000D+01	.1000D+01	.1000D+01	.1000D+01	.1000D+01	.1000D+01
40.0	.1000D+01	.1000D+01	.1000D+01	.1000D+01	.1000D+01	.1000D+01
45.0	.1000D+01	.1000D+01	.1000D+01	.1000D+01	.1000D+01	.1000D+01
50.0	.1000D+01	.1000D+01	.1000D+01	.1000D+01	.1000D+01	.1000D+01

DIMENSIONLESS CONCENTRATION C/C0 FOR

V = .10 D = 1.00 R = 1.0 LAMBDA = .0 ALPHA = .0 T0 = 60.0

T(YEARS)	X = 400.	X = 500.	X = 600.	X = 700.	X = 800.	X = 1000.
.5	.0000D+00	.0000D+00	.0000D+00	.0000D+00	.0000D+00	.0000D+00
1.0	.0000D+00	.0000D+00	.0000D+00	.0000D+00	.0000D+00	.0000D+00
2.0	.0000D+00	.0000D+00	.0000D+00	.0000D+00	.0000D+00	.0000D+00
3.0	.1834D–09	.0000D+00	.0000D+00	.0000D+00	.0000D+00	.0000D+00
4.0	.1019D–05	.2031D–10	.0000D+00	.0000D+00	.0000D+00	.0000D+00
5.0	.1366D–03	.5843D–07	.1765D–11	.0000D+00	.0000D+00	.0000D+00
6.0	.2840D–02	.9272D–05	.3408D–08	.1362D–12	.0000D+00	.0000D+00
7.0	.2045D–01	.2804D–03	.6100D–06	.2012D–09	.0000D+00	.0000D+00
8.0	.7643D–01	.3020D–02	.2471D–04	.3946D–07	.1199D–10	.0000D+00
9.0	.1861D+00	.1641D–01	.3717D–03	.2014D–05	.2525D–08	.0000D+00
10.0	.3394D+00	.5551D–01	.2806D–02	.4013D–04	.1556D–06	.4338D–13
15.0	.9232D+00	.6772D+00	.3073D+00	.7153D–01	.7675D–02	.7117D–05
20.0	.9971D+00	.9728D+00	.8615D+00	.5998D+00	.2811D+00	.1251D–01
25.0	.9999D+00	.9990D+00	.9902D+00	.9437D+00	.7999D+00	.2589D+00
30.0	.1000D+01	.1000D+01	.9996D+00	.9964D+00	.9777D+00	.7419D+00
35.0	.1000D+01	.1000D+01	.1000D+01	.9999D+00	.9987D+00	.9598D+00
40.0	.1000D+01	.1000D+01	.1000D+01	.1000D+01	.9999D+00	.9966D+00
45.0	.1000D+01	.1000D+01	.1000D+01	.1000D+01	.1000D+01	.9998D+00
50.0	.1000D+01	.1000D+01	.1000D+01	.1000D+01	.1000D+01	.1000D+01

DIMENSIONLESS CONCENTRATION C/C0 FOR

V = .10 D = 5.00 R = 1.0 LAMBDA = .0 ALPHA = .0 T0 = 60.0

T(YEARS)	X = 50.	X = 100.	X = 150.	X = 200.	X = 250.	X = 300.
.5	.1334D+00	.1238D-01	.3586D-03	.3039D-05	.7125D-08	.4522D-11
1.0	.3172D+00	.9541D-01	.1668D-01	.1617D-02	.8549D-04	.2390D-05
2.0	.5585D+00	.3181D+00	.1417D+00	.4813D-01	.1224D-01	.2314D-02
3.0	.6999D+00	.4998D+00	.3074D+00	.1599D+00	.6934D-01	.2482D-01
4.0	.7892D+00	.6319D+00	.4571D+00	.2944D+00	.1669D+00	.8250D-01
5.0	.8485D+00	.7270D+00	.5789D+00	.4233D+00	.2814D+00	.1686D+00
6.0	.8893D+00	.7958D+00	.6744D+00	.5356D+00	.3951D+00	.2688D+00
7.0	.9180D+00	.8461D+00	.7481D+00	.6290D+00	.4990D+00	.3709D+00
8.0	.9387D+00	.8832D+00	.8048D+00	.7049D+00	.5893D+00	.4672D+00
9.0	.9537D+00	.9108D+00	.8484D+00	.7658D+00	.6657D+00	.5540D+00
10.0	.9648D+00	.9315D+00	.8819D+00	.8143D+00	.7291D+00	.6298D+00
15.0	.9904D+00	.9806D+00	.9649D+00	.9414D+00	.9081D+00	.8635D+00
20.0	.9971D+00	.9941D+00	.9890D+00	.9810D+00	.9690D+00	.9516D+00
25.0	.9991D+00	.9981D+00	.9964D+00	.9937D+00	.9894D+00	.9829D+00
30.0	.9997D+00	.9994D+00	.9988D+00	.9979D+00	.9963D+00	.9939D+00
35.0	.9999D+00	.9998D+00	.9996D+00	.9993D+00	.9987D+00	.9979D+00
40.0	.1000D+01	.9999D+00	.9999D+00	.9997D+00	.9996D+00	.9992D+00
45.0	.1000D+01	.1000D+01	.1000D+01	.9999D+00	.9998D+00	.9997D+00
50.0	.1000D+01	.1000D+01	.1000D+01	.1000D+01	.9999D+00	.9999D+00

DIMENSIONLESS CONCENTRATION C/C0 FOR

V = .10 D = 5.00 R = 1.0 LAMBDA = .0 ALPHA = .0 T0 = 60.0

T(YEARS)	X = 400.	X = 500.	X = 600.	X = 700.	X = 800.	X = 1000.
.5	.0000D+00	.0000D+00	.0000D+00	.0000D+00	.0000D+00	.0000D+00
1.0	.2679D-09	.0000D+00	.0000D+00	.0000D+00	.0000D+00	.0000D+00
2.0	.3239D-04	.1258D-06	.1323D-09	.3701D-13	.0000D+00	.0000D+00
3.0	.1756D-02	.5398D-04	.7049D-06	.3859D-08	.8762D-11	.0000D+00
4.0	.1311D-01	.1135D-02	.5238D-04	.1269D-05	.1602D-07	.3516D-12
5.0	.4352D-01	.7006D-02	.6889D-03	.4082D-04	.1442D-05	.3736D-09
6.0	.9525D-01	.2324D-01	.3780D-02	.4059D-03	.2850D-04	.3826D-07
7.0	.1649D+00	.5383D-01	.1253D-01	.2056D-02	.2354D-03	.1024D-05
8.0	.2455D+00	.9949D-01	.3024D-01	.6809D-02	.1125D-02	.1180D-04
9.0	.3304D+00	.1580D+00	.5900D-01	.1697D-01	.3723D-02	.7735D-04
10.0	.4143D+00	.2257D+00	.9915D-01	.3464D-01	.9525D-02	.3413D-03
15.0	.7389D+00	.5778D+00	.4033D+00	.2476D+00	.1320D+00	.2377D-01
20.0	.8962D+00	.8086D+00	.6869D+00	.5419D+00	.3921D+00	.1522D+00
25.0	.9609D+00	.9205D+00	.8556D+00	.7633D+00	.6465D+00	.3817D+00
30.0	.9854D+00	.9684D+00	.9380D+00	.8893D+00	.8185D+00	.6138D+00
35.0	.9946D+00	.9877D+00	.9745D+00	.9513D+00	.9141D+00	.7852D+00
40.0	.9980D+00	.9953D+00	.9897D+00	.9794D+00	.9615D+00	.8901D+00
45.0	.9993D+00	.9982D+00	.9959D+00	.9915D+00	.9833D+00	.9470D+00
50.0	.9997D+00	.9993D+00	.9984D+00	.9966D+00	.9930D+00	.9756D+00

DIMENSIONLESS CONCENTRATION C/C0 FOR

V = .20 D = .20 R = 1.0 LAMBDA = .0 ALPHA = .0 T0 = 60.0

T(YEARS)	X = 50.	X = 100.	X = 150.	X = 200.	X = 250.	X = 300.
.5	.5551D–01	.4338D–13	.0000D+00	.0000D+00	.0000D+00	.0000D+00
1.0	.9728D+00	.1251D–01	.8496D–10	.0000D+00	.0000D+00	.0000D+00
2.0	.1000D+01	.9966D+00	.4093D+00	.7858D–03	.5644D–09	.0000D+00
3.0	.1000D+01	.1000D+01	.9995D+00	.8204D+00	.6988D–01	.5482D–04
4.0	.1000D+01	.1000D+01	.1000D+01	.9999D+00	.9598D+00	.3732D+00
5.0	.1000D+01	.1000D+01	.1000D+01	.1000D+01	.1000D+01	.9922D+00
6.0	.1000D+01	.1000D+01	.1000D+01	.1000D+01	.1000D+01	.1000D+01
7.0	.1000D+01	.1000D+01	.1000D+01	.1000D+01	.1000D+01	.1000D+01
8.0	.1000D+01	.1000D+01	.1000D+01	.1000D+01	.1000D+01	.1000D+01
9.0	.1000D+01	.1000D+01	.1000D+01	.1000D+01	.1000D+01	.1000D+01
10.0	.1000D+01	.1000D+01	.1000D+01	.1000D+01	.1000D+01	.1000D+01
15.0	.1000D+01	.1000D+01	.1000D+01	.1000D+01	.1000D+01	.1000D+01
20.0	.1000D+01	.1000D+01	.1000D+01	.1000D+01	.1000D+01	.1000D+01
25.0	.1000D+01	.1000D+01	.1000D+01	.1000D+01	.1000D+01	.1000D+01
30.0	.1000D+01	.1000D+01	.1000D+01	.1000D+01	.1000D+01	.1000D+01
35.0	.1000D+01	.1000D+01	.1000D+01	.1000D+01	.1000D+01	.1000D+01
40.0	.1000D+01	.1000D+01	.1000D+01	.1000D+01	.1000D+01	.1000D+01
45.0	.1000D+01	.1000D+01	.1000D+01	.1000D+01	.1000D+01	.1000D+01
50.0	.1000D+01	.1000D+01	.1000D+01	.1000D+01	.1000D+01	.1000D+01

DIMENSIONLESS CONCENTRATION C/C0 FOR

V = .20 D = .20 R = 1.0 LAMBDA = .0 ALPHA = .0 T0 = 60.0

T(YEARS)	X = 400.	X = 500.	X = 600.	X = 700.	X = 800.	X = 1000.
.5	.0000D+00	.0000D+00	.0000D+00	.0000D+00	.0000D+00	.0000D+00
1.0	.0000D+00	.0000D+00	.0000D+00	.0000D+00	.0000D+00	.0000D+00
2.0	.0000D+00	.0000D+00	.0000D+00	.0000D+00	.0000D+00	.0000D+00
3.0	.0000D+00	.0000D+00	.0000D+00	.0000D+00	.0000D+00	.0000D+00
4.0	.4006D–05	.0000D+00	.0000D+00	.0000D+00	.0000D+00	.0000D+00
5.0	.9898D–01	.3010D–06	.0000D+00	.0000D+00	.0000D+00	.0000D+00
6.0	.9024D+00	.1849D–01	.2302D–07	.0000D+00	.0000D+00	.0000D+00
7.0	.9998D+00	.6388D+00	.2765D–02	.1782D–08	.0000D+00	.0000D+00
8.0	.1000D+01	.9933D+00	.3239D+00	.3575D–03	.1393D–09	.0000D+00
9.0	.1000D+01	.1000D+01	.9436D+00	.1201D+00	.4185D–04	.0000D+00
10.0	.1000D+01	.1000D+01	.9997D+00	.7877D+00	.3442D–01	.8679D–12
15.0	.1000D+01	.1000D+01	.1000D+01	.1000D+01	.1000D+01	.9796D+00
20.0	.1000D+01	.1000D+01	.1000D+01	.1000D+01	.1000D+01	.1000D+01
25.0	.1000D+01	.1000D+01	.1000D+01	.1000D+01	.1000D+01	.1000D+01
30.0	.1000D+01	.1000D+01	.1000D+01	.1000D+01	.1000D+01	.1000D+01
35.0	.1000D+01	.1000D+01	.1000D+01	.1000D+01	.1000D+01	.1000D+01
40.0	.1000D+01	.1000D+01	.1000D+01	.1000D+01	.1000D+01	.1000D+01
45.0	.1000D+01	.1000D+01	.1000D+01	.1000D+01	.1000D+01	.1000D+01
50.0	.1000D+01	.1000D+01	.1000D+01	.1000D+01	.1000D+01	.1000D+01

DIMENSIONLESS CONCENTRATION C/C0 FOR

V = .20 D = 2.00 R = 1.0 LAMBDA = .0 ALPHA = .0 T0 = 60.0

T(YEARS)	X = 50.	X = 100.	X = 150.	X = 200.	X = 250.	X = 300.
.5	.2814D+00	.7006D–02	.8193D–05	.3736D–09	.0000D+00	.0000D+00
1.0	.7291D+00	.2257D+00	.1873D–01	.3413D–03	.1255D–05	.8955D–09
2.0	.9690D+00	.8086D+00	.4663D+00	.1522D+00	.2487D–01	.1906D–02
3.0	.9963D+00	.9684D+00	.8567D+00	.6138D+00	.3157D+00	.1067D+00
4.0	.9996D+00	.9953D+00	.9716D+00	.8901D+00	.7114D+00	.4572D+00
5.0	.9999D+00	.9993D+00	.9950D+00	.9756D+00	.9144D+00	.7798D+00
6.0	.1000D+01	.9999D+00	.9992D+00	.9953D+00	.9795D+00	.9327D+00
7.0	.1000D+01	.1000D+01	.9999D+00	.9991D+00	.9957D+00	.9829D+00
8.0	.1000D+01	.1000D+01	.1000D+01	.9999D+00	.9992D+00	.9962D+00
9.0	.1000D+01	.1000D+01	.1000D+01	.1000D+01	.9998D+00	.9992D+00
10.0	.1000D+01	.1000D+01	.1000D+01	.1000D+01	.1000D+01	.9998D+00
15.0	.1000D+01	.1000D+01	.1000D+01	.1000D+01	.1000D+01	.1000D+01
20.0	.1000D+01	.1000D+01	.1000D+01	.1000D+01	.1000D+01	.1000D+01
25.0	.1000D+01	.1000D+01	.1000D+01	.1000D+01	.1000D+01	.1000D+01
30.0	.1000D+01	.1000D+01	.1000D+01	.1000D+01	.1000D+01	.1000D+01
35.0	.1000D+01	.1000D+01	.1000D+01	.1000D+01	.1000D+01	.1000D+01
40.0	.1000D+01	.1000D+01	.1000D+01	.1000D+01	.1000D+01	.1000D+01
45.0	.1000D+01	.1000D+01	.1000D+01	.1000D+01	.1000D+01	.1000D+01
50.0	.1000D+01	.1000D+01	.1000D+01	.1000D+01	.1000D+01	.1000D+01

DIMENSIONLESS CONCENTRATION C/C0 FOR

V = .20 D = 2.00 R = 1.0 LAMBDA = .0 ALPHA = .0 T0 = 60.0

T(YEARS)	X = 400.	X = 500.	X = 600.	X = 700.	X = 800.	X = 1000.
.5	.0000D+00	.0000D+00	.0000D+00	.0000D+00	.0000D+00	.0000D+00
1.0	.0000D+00	.0000D+00	.0000D+00	.0000D+00	.0000D+00	.0000D+00
2.0	.1019D–05	.2031D–10	.0000D+00	.0000D+00	.0000D+00	.0000D+00
3.0	.2840D–02	.9272D–05	.3408D–08	.1362D–12	.0000D+00	.0000D+00
4.0	.7643D–01	.3020D–02	.2471D–04	.3946D–07	.1199D–10	.0000D+00
5.0	.3394D+00	.5551D–01	.2806D–02	.4013D–04	.1556D–06	.4338D–13
6.0	.6596D+00	.2523D+00	.4072D–01	.2437D–02	.5081D–04	.8342D–09
7.0	.8669D+00	.5444D+00	.1880D+00	.3009D–01	.2037D–02	.6015D–06
8.0	.9576D+00	.7842D+00	.4416D+00	.1405D+00	.2236D–01	.5599D–04
9.0	.9884D+00	.9169D+00	.6927D+00	.3539D+00	.1053D+00	.1340D–02
10.0	.9971D+00	.9728D+00	.8615D+00	.5998D+00	.2811D+00	.1251D–01
15.0	.1000D+01	.1000D+01	.9996D+00	.9964D+00	.9777D+00	.7419D+00
20.0	.1000D+01	.1000D+01	.1000D+01	.1000D+01	.9999D+00	.9966D+00
25.0	.1000D+01	.1000D+01	.1000D+01	.1000D+01	.1000D+01	.1000D+01
30.0	.1000D+01	.1000D+01	.1000D+01	.1000D+01	.1000D+01	.1000D+01
35.0	.1000D+01	.1000D+01	.1000D+01	.1000D+01	.1000D+01	.1000D+01
40.0	.1000D+01	.1000D+01	.1000D+01	.1000D+01	.1000D+01	.1000D+01
45.0	.1000D+01	.1000D+01	.1000D+01	.1000D+01	.1000D+01	.1000D+01
50.0	.1000D+01	.1000D+01	.1000D+01	.1000D+01	.1000D+01	.1000D+01

DIMENSIONLESS CONCENTRATION C/C0 FOR

V = .20 D = 10.00 R = 1.0 LAMBDA = .0 ALPHA = .0 T0 = 60.0

T(YEARS)	X = 50.	X = 100.	X = 150.	X = 200.	X = 250.	X = 300.
.5	.3172D+00	.9541D-01	.1668D-01	.1617D-02	.8549D-04	.2390D-05
1.0	.5585D+00	.3181D+00	.1417D+00	.4813D-01	.1224D-01	.2314D-02
2.0	.7892D+00	.6319D+00	.4571D+00	.2944D+00	.1669D+00	.8250D-01
3.0	.8893D+00	.7958D+00	.6744D+00	.5356D+00	.3951D+00	.2688D+00
4.0	.9387D+00	.8832D+00	.8048D+00	.7049D+00	.5893D+00	.4672D+00
5.0	.9648D+00	.9315D+00	.8819D+00	.8143D+00	.7291D+00	.6298D+00
6.0	.9793D+00	.9591D+00	.9279D+00	.8832D+00	.8236D+00	.7492D+00
7.0	.9876D+00	.9752D+00	.9555D+00	.9263D+00	.8857D+00	.8325D+00
8.0	.9925D+00	.9848D+00	.9723D+00	.9533D+00	.9261D+00	.8889D+00
9.0	.9954D+00	.9906D+00	.9826D+00	.9703D+00	.9521D+00	.9266D+00
10.0	.9971D+00	.9941D+00	.9890D+00	.9810D+00	.9690D+00	.9516D+00
15.0	.9997D+00	.9994D+00	.9988D+00	.9979D+00	.9963D+00	.9939D+00
20.0	.1000D+01	.9999D+00	.9999D+00	.9997D+00	.9996D+00	.9992D+00
25.0	.1000D+01	.1000D+01	.1000D+01	.1000D+01	.9999D+00	.9999D+00
30.0	.1000D+01	.1000D+01	.1000D+01	.1000D+01	.1000D+01	.1000D+01
35.0	.1000D+01	.1000D+01	.1000D+01	.1000D+01	.1000D+01	.1000D+01
40.0	.1000D+01	.1000D+01	.1000D+01	.1000D+01	.1000D+01	.1000D+01
45.0	.1000D+01	.1000D+01	.1000D+01	.1000D+01	.1000D+01	.1000D+01
50.0	.1000D+01	.1000D+01	.1000D+01	.1000D+01	.1000D+01	.1000D+01

DIMENSIONLESS CONCENTRATION C/C0 FOR

V = .20 D = 10.00 R = 1.0 LAMBDA = .0 ALPHA = .0 T0 = 60.0

T(YEARS)	X = 400.	X = 500.	X = 600.	X = 700.	X = 800.	X = 1000.
.5	.2679D-09	.0000D+00	.0000D+00	.0000D+00	.0000D+00	.0000D+00
1.0	.3239D-04	.1258D-06	.1323D-09	.3701D-13	.0000D+00	.0000D+00
2.0	.1311D-01	.1135D-02	.5238D-04	.1269D-05	.1602D-07	.3516D-12
3.0	.9525D-01	.2324D-01	.3780D-02	.4059D-03	.2850D-04	.3826D-07
4.0	.2455D+00	.9949D-01	.3024D-01	.6809D-02	.1125D-02	.1180D-04
5.0	.4143D+00	.2257D+00	.9915D-01	.3464D-01	.9525D-02	.3413D-03
6.0	.5666D+00	.3726D+00	.2077D+00	.9670D-01	.3718D-01	.3004D-02
7.0	.6892D+00	.5140D+00	.3371D+00	.1916D+00	.9317D-01	.1337D-01
8.0	.7816D+00	.6357D+00	.4679D+00	.3068D+00	.1771D+00	.3881D-01
9.0	.8487D+00	.7336D+00	.5866D+00	.4276D+00	.2806D+00	.8483D-01
10.0	.8962D+00	.8086D+00	.6869D+00	.5419D+00	.3921D+00	.1522D+00
15.0	.9854D+00	.9684D+00	.9380D+00	.8893D+00	.8185D+00	.6138D+00
20.0	.9980D+00	.9953D+00	.9897D+00	.9794D+00	.9615D+00	.8901D+00
25.0	.9997D+00	.9993D+00	.9984D+00	.9966D+00	.9930D+00	.9756D+00
30.0	.1000D+01	.9999D+00	.9998D+00	.9995D+00	.9988D+00	.9953D+00
35.0	.1000D+01	.1000D+01	.1000D+01	.9999D+00	.9998D+00	.9991D+00
40.0	.1000D+01	.1000D+01	.1000D+01	.1000D+01	.1000D+01	.9999D+00
45.0	.1000D+01	.1000D+01	.1000D+01	.1000D+01	.1000D+01	.1000D+01
50.0	.1000D+01	.1000D+01	.1000D+01	.1000D+01	.1000D+01	.1000D+01

DIMENSIONLESS CONCENTRATION C/C0 FOR

V = .40 D = .40 R = 1.0 LAMBDA = .0 ALPHA = .0 T0 = 60.0

T(YEARS)	X = 50.	X = 100.	X = 150.	X = 200.	X = 250.	X = 300.
.5	.9728D+00	.1251D–01	.8496D–10	.0000D+00	.0000D+00	.0000D+00
1.0	.1000D+01	.9966D+00	.4093D+00	.7858D–03	.5644D–09	.0000D+00
2.0	.1000D+01	.1000D+01	.1000D+01	.9999D+00	.9598D+00	.3732D+00
3.0	.1000D+01	.1000D+01	.1000D+01	.1000D+01	.1000D+01	.1000D+01
4.0	.1000D+01	.1000D+01	.1000D+01	.1000D+01	.1000D+01	.1000D+01
5.0	.1000D+01	.1000D+01	.1000D+01	.1000D+01	.1000D+01	.1000D+01
6.0	.1000D+01	.1000D+01	.1000D+01	.1000D+01	.1000D+01	.1000D+01
7.0	.1000D+01	.1000D+01	.1000D+01	.1000D+01	.1000D+01	.1000D+01
8.0	.1000D+01	.1000D+01	.1000D+01	.1000D+01	.1000D+01	.1000D+01
9.0	.1000D+01	.1000D+01	.1000D+01	.1000D+01	.1000D+01	.1000D+01
10.0	.1000D+01	.1000D+01	.1000D+01	.1000D+01	.1000D+01	.1000D+01
15.0	.1000D+01	.1000D+01	.1000D+01	.1000D+01	.1000D+01	.1000D+01
20.0	.1000D+01	.1000D+01	.1000D+01	.1000D+01	.1000D+01	.1000D+01
25.0	.1000D+01	.1000D+01	.1000D+01	.1000D+01	.1000D+01	.1000D+01
30.0	.1000D+01	.1000D+01	.1000D+01	.1000D+01	.1000D+01	.1000D+01
35.0	.1000D+01	.1000D+01	.1000D+01	.1000D+01	.1000D+01	.1000D+01
40.0	.1000D+01	.1000D+01	.1000D+01	.1000D+01	.1000D+01	.1000D+01
45.0	.1000D+01	.1000D+01	.1000D+01	.1000D+01	.1000D+01	.1000D+01
50.0	.1000D+01	.1000D+01	.1000D+01	.1000D+01	.1000D+01	.1000D+01

DIMENSIONLESS CONCENTRATION C/C0 FOR

V = .40 D = .40 R = 1.0 LAMBDA = .0 ALPHA = .0 T0 = 60.0

T(YEARS)	X = 400.	X = 500.	X = 600.	X = 700.	X = 800.	X = 1000.
.5	.0000D+00	.0000D+00	.0000D+00	.0000D+00	.0000D+00	.0000D+00
1.0	.0000D+00	.0000D+00	.0000D+00	.0000D+00	.0000D+00	.0000D+00
2.0	.4006D–05	.0000D+00	.0000D+00	.0000D+00	.0000D+00	.0000D+00
3.0	.9024D+00	.1849D–01	.2302D–07	.0000D+00	.0000D+00	.0000D+00
4.0	.1000D+01	.9933D+00	.3239D+00	.3575D–03	.1393D–09	.0000D+00
5.0	.1000D+01	.1000D+01	.9997D+00	.7877D+00	.3442D–01	.8679D–12
6.0	.1000D+01	.1000D+01	.1000D+01	.1000D+01	.9664D+00	.1596D–02
7.0	.1000D+01	.1000D+01	.1000D+01	.1000D+01	.1000D+01	.6922D+00
8.0	.1000D+01	.1000D+01	.1000D+01	.1000D+01	.1000D+01	.9998D+00
9.0	.1000D+01	.1000D+01	.1000D+01	.1000D+01	.1000D+01	.1000D+01
10.0	.1000D+01	.1000D+01	.1000D+01	.1000D+01	.1000D+01	.1000D+01
15.0	.1000D+01	.1000D+01	.1000D+01	.1000D+01	.1000D+01	.1000D+01
20.0	.1000D+01	.1000D+01	.1000D+01	.1000D+01	.1000D+01	.1000D+01
25.0	.1000D+01	.1000D+01	.1000D+01	.1000D+01	.1000D+01	.1000D+01
30.0	.1000D+01	.1000D+01	.1000D+01	.1000D+01	.1000D+01	.1000D+01
35.0	.1000D+01	.1000D+01	.1000D+01	.1000D+01	.1000D+01	.1000D+01
40.0	.1000D+01	.1000D+01	.1000D+01	.1000D+01	.1000D+01	.1000D+01
45.0	.1000D+01	.1000D+01	.1000D+01	.1000D+01	.1000D+01	.1000D+01
50.0	.1000D+01	.1000D+01	.1000D+01	.1000D+01	.1000D+01	.1000D+01

DIMENSIONLESS CONCENTRATION C/C0 FOR

V = .40 D = 4.00 R = 1.0 LAMBDA = .0 ALPHA = .0 T0 = 60.0

T(YEARS)	X = 50.	X = 100.	X = 150.	X = 200.	X = 250.	X = 300.
.5	.7291D+00	.2257D+00	.1873D–01	.3413D–03	.1255D–05	.8955D–09
1.0	.9690D+00	.8086D+00	.4663D+00	.1522D+00	.2487D–01	.1906D–02
2.0	.9996D+00	.9953D+00	.9716D+00	.8901D+00	.7114D+00	.4572D+00
3.0	.1000D+01	.9999D+00	.9992D+00	.9953D+00	.9795D+00	.9327D+00
4.0	.1000D+01	.1000D+01	.1000D+01	.9999D+00	.9992D+00	.9962D+00
5.0	.1000D+01	.1000D+01	.1000D+01	.1000D+01	.1000D+01	.9998D+00
6.0	.1000D+01	.1000D+01	.1000D+01	.1000D+01	.1000D+01	.1000D+01
7.0	.1000D+01	.1000D+01	.1000D+01	.1000D+01	.1000D+01	.1000D+01
8.0	.1000D+01	.1000D+01	.1000D+01	.1000D+01	.1000D+01	.1000D+01
9.0	.1000D+01	.1000D+01	.1000D+01	.1000D+01	.1000D+01	.1000D+01
10.0	.1000D+01	.1000D+01	.1000D+01	.1000D+01	.1000D+01	.1000D+01
15.0	.1000D+01	.1000D+01	.1000D+01	.1000D+01	.1000D+01	.1000D+01
20.0	.1000D+01	.1000D+01	.1000D+01	.1000D+01	.1000D+01	.1000D+01
25.0	.1000D+01	.1000D+01	.1000D+01	.1000D+01	.1000D+01	.1000D+01
30.0	.1000D+01	.1000D+01	.1000D+01	.1000D+01	.1000D+01	.1000D+01
35.0	.1000D+01	.1000D+01	.1000D+01	.1000D+01	.1000D+01	.1000D+01
40.0	.1000D+01	.1000D+01	.1000D+01	.1000D+01	.1000D+01	.1000D+01
45.0	.1000D+01	.1000D+01	.1000D+01	.1000D+01	.1000D+01	.1000D+01
50.0	.1000D+01	.1000D+01	.1000D+01	.1000D+01	.1000D+01	.1000D+01

DIMENSIONLESS CONCENTRATION C/C0 FOR

V = .40 D = 4.00 R = 1.0 LAMBDA = .0 ALPHA = .0 T0 = 60.0

T(YEARS)	X = 400.	X = 500.	X = 600.	X = 700.	X = 800.	X = 1000.
.5	.0000D+00	.0000D+00	.0000D+00	.0000D+00	.0000D+00	.0000D+00
1.0	.1019D–05	.2031D–10	.0000D+00	.0000D+00	.0000D+00	.0000D+00
2.0	.7643D–01	.3020D–02	.2471D–04	.3946D–07	.1199D–10	.0000D+00
3.0	.6596D+00	.2523D+00	.4072D–01	.2437D–02	.5081D–04	.8342D–09
4.0	.9576D+00	.7842D+00	.4416D+00	.1405D+00	.2236D–01	.5599D–04
5.0	.9971D+00	.9728D+00	.8615D+00	.5998D+00	.2811D+00	.1251D–01
6.0	.9999D+00	.9979D+00	.9823D+00	.9102D+00	.7194D+00	.1743D+00
7.0	.1000D+01	.9999D+00	.9985D+00	.9884D+00	.9413D+00	.5631D+00
8.0	.1000D+01	.1000D+01	.9999D+00	.9990D+00	.9924D+00	.8662D+00
9.0	.1000D+01	.1000D+01	.1000D+01	.9999D+00	.9993D+00	.9744D+00
10.0	.1000D+01	.1000D+01	.1000D+01	.1000D+01	.9999D+00	.9966D+00
15.0	.1000D+01	.1000D+01	.1000D+01	.1000D+01	.1000D+01	.1000D+01
20.0	.1000D+01	.1000D+01	.1000D+01	.1000D+01	.1000D+01	.1000D+01
25.0	.1000D+01	.1000D+01	.1000D+01	.1000D+01	.1000D+01	.1000D+01
30.0	.1000D+01	.1000D+01	.1000D+01	.1000D+01	.1000D+01	.1000D+01
35.0	.1000D+01	.1000D+01	.1000D+01	.1000D+01	.1000D+01	.1000D+01
40.0	.1000D+01	.1000D+01	.1000D+01	.1000D+01	.1000D+01	.1000D+01
45.0	.1000D+01	.1000D+01	.1000D+01	.1000D+01	.1000D+01	.1000D+01
50.0	.1000D+01	.1000D+01	.1000D+01	.1000D+01	.1000D+01	.1000D+01

DIMENSIONLESS CONCENTRATION C/C0 FOR

V = .40 D = 20.00 R = 1.0 LAMBDA = .0 ALPHA = .0 T0 = 60.0

T(YEARS)	X = 50.	X = 100.	X = 150.	X = 200.	X = 250.	X = 300.
.5	.5585D+00	.3181D+00	.1417D+00	.4813D–01	.1224D–01	.2314D–02
1.0	.7892D+00	.6319D+00	.4571D+00	.2944D+00	.1669D+00	.8250D–01
2.0	.9387D+00	.8832D+00	.8048D+00	.7049D+00	.5893D+00	.4672D+00
3.0	.9793D+00	.9591D+00	.9279D+00	.8832D+00	.8236D+00	.7492D+00
4.0	.9925D+00	.9848D+00	.9723D+00	.9533D+00	.9261D+00	.8889D+00
5.0	.9971D+00	.9941D+00	.9890D+00	.9810D+00	.9690D+00	.9516D+00
6.0	.9989D+00	.9976D+00	.9956D+00	.9922D+00	.9869D+00	.9790D+00
7.0	.9995D+00	.9990D+00	.9982D+00	.9967D+00	.9944D+00	.9908D+00
8.0	.9998D+00	.9996D+00	.9992D+00	.9986D+00	.9976D+00	.9961D+00
9.0	.9999D+00	.9998D+00	.9997D+00	.9994D+00	.9990D+00	.9983D+00
10.0	.1000D+01	.9999D+00	.9999D+00	.9997D+00	.9996D+00	.9992D+00
15.0	.1000D+01	.1000D+01	.1000D+01	.1000D+01	.1000D+01	.1000D+01
20.0	.1000D+01	.1000D+01	.1000D+01	.1000D+01	.1000D+01	.1000D+01
25.0	.1000D+01	.1000D+01	.1000D+01	.1000D+01	.1000D+01	.1000D+01
30.0	.1000D+01	.1000D+01	.1000D+01	.1000D+01	.1000D+01	.1000D+01
35.0	.1000D+01	.1000D+01	.1000D+01	.1000D+01	.1000D+01	.1000D+01
40.0	.1000D+01	.1000D+01	.1000D+01	.1000D+01	.1000D+01	.1000D+01
45.0	.1000D+01	.1000D+01	.1000D+01	.1000D+01	.1000D+01	.1000D+01
50.0	.1000D+01	.1000D+01	.1000D+01	.1000D+01	.1000D+01	.1000D+01

DIMENSIONLESS CONCENTRATION C/C0 FOR

V = .40 D = 20.00 R = 1.0 LAMBDA = .0 ALPHA = .0 T0 = 60.0

T(YEARS)	X = 400.	X = 500.	X = 600.	X = 700.	X = 800.	X = 1000.
.5	.3239D–04	.1258D–06	.1323D–09	.3701D–13	.0000D+00	.0000D+00
1.0	.1311D–01	.1135D–02	.5238D–04	.1269D–05	.1602D–07	.3516D–12
2.0	.2455D+00	.9949D–01	.3024D–01	.6809D–02	.1125D–02	.1180D–04
3.0	.5666D+00	.3726D+00	.2077D+00	.9670D–01	.3718D–01	.3004D–02
4.0	.7816D+00	.6357D+00	.4679D+00	.3068D+00	.1771D+00	.3881D–01
5.0	.8962D+00	.8086D+00	.6869D+00	.5419D+00	.3921D+00	.1522D+00
6.0	.9524D+00	.9048D+00	.8302D+00	.7273D+00	.6013D+00	.3324D+00
7.0	.9784D+00	.9541D+00	.9125D+00	.8486D+00	.7602D+00	.5262D+00
8.0	.9902D+00	.9783D+00	.9564D+00	.9198D+00	.8643D+00	.6909D+00
9.0	.9956D+00	.9899D+00	.9787D+00	.9589D+00	.9265D+00	.8110D+00
10.0	.9980D+00	.9953D+00	.9897D+00	.9794D+00	.9615D+00	.8901D+00
15.0	.1000D+01	.9999D+00	.9998D+00	.9995D+00	.9988D+00	.9953D+00
20.0	.1000D+01	.1000D+01	.1000D+01	.1000D+01	.1000D+01	.9999D+00
25.0	.1000D+01	.1000D+01	.1000D+01	.1000D+01	.1000D+01	.1000D+01
30.0	.1000D+01	.1000D+01	.1000D+01	.1000D+01	.1000D+01	.1000D+01
35.0	.1000D+01	.1000D+01	.1000D+01	.1000D+01	.1000D+01	.1000D+01
40.0	.1000D+01	.1000D+01	.1000D+01	.1000D+01	.1000D+01	.1000D+01
45.0	.1000D+01	.1000D+01	.1000D+01	.1000D+01	.1000D+01	.1000D+01
50.0	.1000D+01	.1000D+01	.1000D+01	.1000D+01	.1000D+01	.1000D+01

DIMENSIONLESS CONCENTRATION C/C0 FOR

V = .60 D = .60 R = 1.0 LAMBDA = .0 ALPHA = .0 T0 = 60.0

T(YEARS)	X = 50.	X = 100.	X = 150.	X = 200.	X = 250.	X = 300.
.5	.1000D+01	.7419D+00	.3072D–02	.4592D–09	.0000D+00	.0000D+00
1.0	.1000D+01	.1000D+01	.9995D+00	.8204D+00	.6988D–01	.5482D–04
2.0	.1000D+01	.1000D+01	.1000D+01	.1000D+01	.1000D+01	.1000D+01
3.0	.1000D+01	.1000D+01	.1000D+01	.1000D+01	.1000D+01	.1000D+01
4.0	.1000D+01	.1000D+01	.1000D+01	.1000D+01	.1000D+01	.1000D+01
5.0	.1000D+01	.1000D+01	.1000D+01	.1000D+01	.1000D+01	.1000D+01
6.0	.1000D+01	.1000D+01	.1000D+01	.1000D+01	.1000D+01	.1000D+01
7.0	.1000D+01	.1000D+01	.1000D+01	.1000D+01	.1000D+01	.1000D+01
8.0	.1000D+01	.1000D+01	.1000D+01	.1000D+01	.1000D+01	.1000D+01
9.0	.1000D+01	.1000D+01	.1000D+01	.1000D+01	.1000D+01	.1000D+01
10.0	.1000D+01	.1000D+01	.1000D+01	.1000D+01	.1000D+01	.1000D+01
15.0	.1000D+01	.1000D+01	.1000D+01	.1000D+01	.1000D+01	.1000D+01
20.0	.1000D+01	.1000D+01	.1000D+01	.1000D+01	.1000D+01	.1000D+01
25.0	.1000D+01	.1000D+01	.1000D+01	.1000D+01	.1000D+01	.1000D+01
30.0	.1000D+01	.1000D+01	.1000D+01	.1000D+01	.1000D+01	.1000D+01
35.0	.1000D+01	.1000D+01	.1000D+01	.1000D+01	.1000D+01	.1000D+01
40.0	.1000D+01	.1000D+01	.1000D+01	.1000D+01	.1000D+01	.1000D+01
45.0	.1000D+01	.1000D+01	.1000D+01	.1000D+01	.1000D+01	.1000D+01
50.0	.1000D+01	.1000D+01	.1000D+01	.1000D+01	.1000D+01	.1000D+01

DIMENSIONLESS CONCENTRATION C/C0 FOR

V = .60 D = .60 R = 1.0 LAMBDA = .0 ALPHA = .0 T0 = 60.0

T(YEARS)	X = 400.	X = 500.	X = 600.	X = 700.	X = 800.	X = 1000.
.5	.0000D+00	.0000D+00	.0000D+00	.0000D+00	.0000D+00	.0000D+00
1.0	.0000D+00	.0000D+00	.0000D+00	.0000D+00	.0000D+00	.0000D+00
2.0	.9024D+00	.1849D–01	.2302D–07	.0000D+00	.0000D+00	.0000D+00
3.0	.1000D+01	.1000D+01	.9436D+00	.1201D+00	.4185D–04	.0000D+00
4.0	.1000D+01	.1000D+01	.1000D+01	.1000D+01	.9664D+00	.1596D–02
5.0	.1000D+01	.1000D+01	.1000D+01	.1000D+01	.1000D+01	.9796D+00
6.0	.1000D+01	.1000D+01	.1000D+01	.1000D+01	.1000D+01	.1000D+01
7.0	.1000D+01	.1000D+01	.1000D+01	.1000D+01	.1000D+01	.1000D+01
8.0	.1000D+01	.1000D+01	.1000D+01	.1000D+01	.1000D+01	.1000D+01
9.0	.1000D+01	.1000D+01	.1000D+01	.1000D+01	.1000D+01	.1000D+01
10.0	.1000D+01	.1000D+01	.1000D+01	.1000D+01	.1000D+01	.1000D+01
15.0	.1000D+01	.1000D+01	.1000D+01	.1000D+01	.1000D+01	.1000D+01
20.0	.1000D+01	.1000D+01	.1000D+01	.1000D+01	.1000D+01	.1000D+01
25.0	.1000D+01	.1000D+01	.1000D+01	.1000D+01	.1000D+01	.1000D+01
30.0	.1000D+01	.1000D+01	.1000D+01	.1000D+01	.1000D+01	.1000D+01
35.0	.1000D+01	.1000D+01	.1000D+01	.1000D+01	.1000D+01	.1000D+01
40.0	.1000D+01	.1000D+01	.1000D+01	.1000D+01	.1000D+01	.1000D+01
45.0	.1000D+01	.1000D+01	.1000D+01	.1000D+01	.1000D+01	.1000D+01
50.0	.1000D+01	.1000D+01	.1000D+01	.1000D+01	.1000D+01	.1000D+01

DIMENSIONLESS CONCENTRATION C/C0 FOR

V = .60 D = 6.00 R = 1.0 LAMBDA = .0 ALPHA = .0 T0 = 60.0

T(YEARS)	X = 50.	X = 100.	X = 150.	X = 200.	X = 250.	X = 300.
.5	.9081D+00	.5778D+00	.1841D+00	.2377D–01	.1119D–02	.1825D–04
1.0	.9963D+00	.9684D+00	.8567D+00	.6138D+00	.3157D+00	.1067D+00
2.0	.1000D+01	.9999D+00	.9992D+00	.9953D+00	.9795D+00	.9327D+00
3.0	.1000D+01	.1000D+01	.1000D+01	.1000D+01	.9998D+00	.9992D+00
4.0	.1000D+01	.1000D+01	.1000D+01	.1000D+01	.1000D+01	.1000D+01
5.0	.1000D+01	.1000D+01	.1000D+01	.1000D+01	.1000D+01	.1000D+01
6.0	.1000D+01	.1000D+01	.1000D+01	.1000D+01	.1000D+01	.1000D+01
7.0	.1000D+01	.1000D+01	.1000D+01	.1000D+01	.1000D+01	.1000D+01
8.0	.1000D+01	.1000D+01	.1000D+01	.1000D+01	.1000D+01	.1000D+01
9.0	.1000D+01	.1000D+01	.1000D+01	.1000D+01	.1000D+01	.1000D+01
10.0	.1000D+01	.1000D+01	.1000D+01	.1000D+01	.1000D+01	.1000D+01
15.0	.1000D+01	.1000D+01	.1000D+01	.1000D+01	.1000D+01	.1000D+01
20.0	.1000D+01	.1000D+01	.1000D+01	.1000D+01	.1000D+01	.1000D+01
25.0	.1000D+01	.1000D+01	.1000D+01	.1000D+01	.1000D+01	.1000D+01
30.0	.1000D+01	.1000D+01	.1000D+01	.1000D+01	.1000D+01	.1000D+01
35.0	.1000D+01	.1000D+01	.1000D+01	.1000D+01	.1000D+01	.1000D+01
40.0	.1000D+01	.1000D+01	.1000D+01	.1000D+01	.1000D+01	.1000D+01
45.0	.1000D+01	.1000D+01	.1000D+01	.1000D+01	.1000D+01	.1000D+01
50.0	.1000D+01	.1000D+01	.1000D+01	.1000D+01	.1000D+01	.1000D+01

DIMENSIONLESS CONCENTRATION C/C0 FOR

V = .60 D = 6.00 R = 1.0 LAMBDA = .0 ALPHA = .0 T0 = 60.0

T(YEARS)	X = 400.	X = 500.	X = 600.	X = 700.	X = 800.	X = 1000.
.5	.1834D–09	.0000D+00	.0000D+00	.0000D+00	.0000D+00	.0000D+00
1.0	.2840D–02	.9272D–05	.3408D–08	.1362D–12	.0000D+00	.0000D+00
2.0	.6596D+00	.2523D+00	.4072D–01	.2437D–02	.5081D–04	.8342D–09
3.0	.9884D+00	.9169D+00	.6927D+00	.3539D+00	.1053D+00	.1340D–02
4.0	.9999D+00	.9979D+00	.9823D+00	.9102D+00	.7194D+00	.1743D+00
5.0	.1000D+01	.1000D+01	.9996D+00	.9964D+00	.9777D+00	.7419D+00
6.0	.1000D+01	.1000D+01	.1000D+01	.9999D+00	.9993D+00	.9744D+00
7.0	.1000D+01	.1000D+01	.1000D+01	.1000D+01	.1000D+01	.9989D+00
8.0	.1000D+01	.1000D+01	.1000D+01	.1000D+01	.1000D+01	.1000D+01
9.0	.1000D+01	.1000D+01	.1000D+01	.1000D+01	.1000D+01	.1000D+01
10.0	.1000D+01	.1000D+01	.1000D+01	.1000D+01	.1000D+01	.1000D+01
15.0	.1000D+01	.1000D+01	.1000D+01	.1000D+01	.1000D+01	.1000D+01
20.0	.1000D+01	.1000D+01	.1000D+01	.1000D+01	.1000D+01	.1000D+01
25.0	.1000D+01	.1000D+01	.1000D+01	.1000D+01	.1000D+01	.1000D+01
30.0	.1000D+01	.1000D+01	.1000D+01	.1000D+01	.1000D+01	.1000D+01
35.0	.1000D+01	.1000D+01	.1000D+01	.1000D+01	.1000D+01	.1000D+01
40.0	.1000D+01	.1000D+01	.1000D+01	.1000D+01	.1000D+01	.1000D+01
45.0	.1000D+01	.1000D+01	.1000D+01	.1000D+01	.1000D+01	.1000D+01
50.0	.1000D+01	.1000D+01	.1000D+01	.1000D+01	.1000D+01	.1000D+01

DIMENSIONLESS CONCENTRATION C/C0 FOR

V = .60 D = 30.00 R = 1.0 LAMBDA = .0 ALPHA = .0 T0 = 60.0

T(YEARS)	X = 50.	X = 100.	X = 150.	X = 200.	X = 250.	X = 300.
.5	.6999D+00	.4998D+00	.3074D+00	.1599D+00	.6934D–01	.2482D–01
1.0	.8893D+00	.7958D+00	.6744D+00	.5356D+00	.3951D+00	.2688D+00
2.0	.9793D+00	.9591D+00	.9279D+00	.8832D+00	.8236D+00	.7492D+00
3.0	.9954D+00	.9906D+00	.9826D+00	.9703D+00	.9521D+00	.9266D+00
4.0	.9989D+00	.9976D+00	.9956D+00	.9922D+00	.9869D+00	.9790D+00
5.0	.9997D+00	.9994D+00	.9988D+00	.9979D+00	.9963D+00	.9939D+00
6.0	.9999D+00	.9998D+00	.9997D+00	.9994D+00	.9990D+00	.9983D+00
7.0	.1000D+01	.1000D+01	.9999D+00	.9998D+00	.9997D+00	.9995D+00
8.0	.1000D+01	.1000D+01	.1000D+01	.1000D+01	.9999D+00	.9999D+00
9.0	.1000D+01	.1000D+01	.1000D+01	.1000D+01	.1000D+01	.1000D+01
10.0	.1000D+01	.1000D+01	.1000D+01	.1000D+01	.1000D+01	.1000D+01
15.0	.1000D+01	.1000D+01	.1000D+01	.1000D+01	.1000D+01	.1000D+01
20.0	.1000D+01	.1000D+01	.1000D+01	.1000D+01	.1000D+01	.1000D+01
25.0	.1000D+01	.1000D+01	.1000D+01	.1000D+01	.1000D+01	.1000D+01
30.0	.1000D+01	.1000D+01	.1000D+01	.1000D+01	.1000D+01	.1000D+01
35.0	.1000D+01	.1000D+01	.1000D+01	.1000D+01	.1000D+01	.1000D+01
40.0	.1000D+01	.1000D+01	.1000D+01	.1000D+01	.1000D+01	.1000D+01
45.0	.1000D+01	.1000D+01	.1000D+01	.1000D+01	.1000D+01	.1000D+01
50.0	.1000D+01	.1000D+01	.1000D+01	.1000D+01	.1000D+01	.1000D+01

DIMENSIONLESS CONCENTRATION C/C0 FOR

V = .60 D = 30.00 R = 1.0 LAMBDA = .0 ALPHA = .0 T0 = 60.0

T(YEARS)	X = 400.	X = 500.	X = 600.	X = 700.	X = 800.	X = 1000.
.5	.1756D–02	.5398D–04	.7049D–06	.3859D–08	.8762D–11	.0000D+00
1.0	.9525D–01	.2324D–01	.3780D–02	.4059D–03	.2850D–04	.3826D–07
2.0	.5666D+00	.3726D+00	.2077D+00	.9670D–01	.3718D–01	.3004D–02
3.0	.8487D+00	.7336D+00	.5866D+00	.4276D+00	.2806D+00	.8483D–01
4.0	.9524D+00	.9048D+00	.8302D+00	.7273D+00	.6013D+00	.3324D+00
5.0	.9854D+00	.9684D+00	.9380D+00	.8893D+00	.8185D+00	.6138D+00
6.0	.9956D+00	.9899D+00	.9787D+00	.9589D+00	.9265D+00	.8110D+00
7.0	.9987D+00	.9968D+00	.9929D+00	.9855D+00	.9724D+00	.9174D+00
8.0	.9996D+00	.9990D+00	.9977D+00	.9951D+00	.9901D+00	.9666D+00
9.0	.9999D+00	.9997D+00	.9993D+00	.9983D+00	.9965D+00	.9872D+00
10.0	.1000D+01	.9999D+00	.9998D+00	.9995D+00	.9988D+00	.9953D+00
15.0	.1000D+01	.1000D+01	.1000D+01	.1000D+01	.1000D+01	.1000D+01
20.0	.1000D+01	.1000D+01	.1000D+01	.1000D+01	.1000D+01	.1000D+01
25.0	.1000D+01	.1000D+01	.1000D+01	.1000D+01	.1000D+01	.1000D+01
30.0	.1000D+01	.1000D+01	.1000D+01	.1000D+01	.1000D+01	.1000D+01
35.0	.1000D+01	.1000D+01	.1000D+01	.1000D+01	.1000D+01	.1000D+01
40.0	.1000D+01	.1000D+01	.1000D+01	.1000D+01	.1000D+01	.1000D+01
45.0	.1000D+01	.1000D+01	.1000D+01	.1000D+01	.1000D+01	.1000D+01
50.0	.1000D+01	.1000D+01	.1000D+01	.1000D+01	.1000D+01	.1000D+01

DIMENSIONLESS CONCENTRATION C/C0 FOR

V = .80 D = .80 R = 1.0 LAMBDA = .0 ALPHA = .0 T0 = 60.0

T(YEARS)	X = 50.	X = 100.	X = 150.	X = 200.	X = 250.	X = 300.
.5	.1000D+01	.9966D+00	.4093D+00	.7858D–03	.5644D–09	.0000D+00
1.0	.1000D+01	.1000D+01	.1000D+01	.9999D+00	.9598D+00	.3732D+00
2.0	.1000D+01	.1000D+01	.1000D+01	.1000D+01	.1000D+01	.1000D+01
3.0	.1000D+01	.1000D+01	.1000D+01	.1000D+01	.1000D+01	.1000D+01
4.0	.1000D+01	.1000D+01	.1000D+01	.1000D+01	.1000D+01	.1000D+01
5.0	.1000D+01	.1000D+01	.1000D+01	.1000D+01	.1000D+01	.1000D+01
6.0	.1000D+01	.1000D+01	.1000D+01	.1000D+01	.1000D+01	.1000D+01
7.0	.1000D+01	.1000D+01	.1000D+01	.1000D+01	.1000D+01	.1000D+01
8.0	.1000D+01	.1000D+01	.1000D+01	.1000D+01	.1000D+01	.1000D+01
9.0	.1000D+01	.1000D+01	.1000D+01	.1000D+01	.1000D+01	.1000D+01
10.0	.1000D+01	.1000D+01	.1000D+01	.1000D+01	.1000D+01	.1000D+01
15.0	.1000D+01	.1000D+01	.1000D+01	.1000D+01	.1000D+01	.1000D+01
20.0	.1000D+01	.1000D+01	.1000D+01	.1000D+01	.1000D+01	.1000D+01
25.0	.1000D+01	.1000D+01	.1000D+01	.1000D+01	.1000D+01	.1000D+01
30.0	.1000D+01	.1000D+01	.1000D+01	.1000D+01	.1000D+01	.1000D+01
35.0	.1000D+01	.1000D+01	.1000D+01	.1000D+01	.1000D+01	.1000D+01
40.0	.1000D+01	.1000D+01	.1000D+01	.1000D+01	.1000D+01	.1000D+01
45.0	.1000D+01	.1000D+01	.1000D+01	.1000D+01	.1000D+01	.1000D+01
50.0	.1000D+01	.1000D+01	.1000D+01	.1000D+01	.1000D+01	.1000D+01

DIMENSIONLESS CONCENTRATION C/C0 FOR

V = .80 D = .80 R = 1.0 LAMBDA = .0 ALPHA = .0 T0 = 60.0

T(YEARS)	X = 400.	X = 500.	X = 600.	X = 700.	X = 800.	X = 1000.
.5	.0000D+00	.0000D+00	.0000D+00	.0000D+00	.0000D+00	.0000D+00
1.0	.4006D–05	.0000D+00	.0000D+00	.0000D+00	.0000D+00	.0000D+00
2.0	.1000D+01	.9933D+00	.3239D+00	.3575D–03	.1393D–09	.0000D+00
3.0	.1000D+01	.1000D+01	.1000D+01	.1000D+01	.9664D+00	.1596D–02
4.0	.1000D+01	.1000D+01	.1000D+01	.1000D+01	.1000D+01	.9998D+00
5.0	.1000D+01	.1000D+01	.1000D+01	.1000D+01	.1000D+01	.1000D+01
6.0	.1000D+01	.1000D+01	.1000D+01	.1000D+01	.1000D+01	.1000D+01
7.0	.1000D+01	.1000D+01	.1000D+01	.1000D+01	.1000D+01	.1000D+01
8.0	.1000D+01	.1000D+01	.1000D+01	.1000D+01	.1000D+01	.1000D+01
9.0	.1000D+01	.1000D+01	.1000D+01	.1000D+01	.1000D+01	.1000D+01
10.0	.1000D+01	.1000D+01	.1000D+01	.1000D+01	.1000D+01	.1000D+01
15.0	.1000D+01	.1000D+01	.1000D+01	.1000D+01	.1000D+01	.1000D+01
20.0	.1000D+01	.1000D+01	.1000D+01	.1000D+01	.1000D+01	.1000D+01
25.0	.1000D+01	.1000D+01	.1000D+01	.1000D+01	.1000D+01	.1000D+01
30.0	.1000D+01	.1000D+01	.1000D+01	.1000D+01	.1000D+01	.1000D+01
35.0	.1000D+01	.1000D+01	.1000D+01	.1000D+01	.1000D+01	.1000D+01
40.0	.1000D+01	.1000D+01	.1000D+01	.1000D+01	.1000D+01	.1000D+01
45.0	.1000D+01	.1000D+01	.1000D+01	.1000D+01	.1000D+01	.1000D+01
50.0	.1000D+01	.1000D+01	.1000D+01	.1000D+01	.1000D+01	.1000D+01

DIMENSIONLESS CONCENTRATION C/C0 FOR

V = .80 D = 8.00 R = 1.0 LAMBDA = .0 ALPHA = .0 T0 = 60.0

T(YEARS)	X = 50.	X = 100.	X = 150.	X = 200.	X = 250.	X = 300.
.5	.9690D+00	.8086D+00	.4663D+00	.1522D+00	.2487D–01	.1906D–02
1.0	.9996D+00	.9953D+00	.9716D+00	.8901D+00	.7114D+00	.4572D+00
2.0	.1000D+01	.1000D+01	.1000D+01	.9999D+00	.9992D+00	.9962D+00
3.0	.1000D+01	.1000D+01	.1000D+01	.1000D+01	.1000D+01	.1000D+01
4.0	.1000D+01	.1000D+01	.1000D+01	.1000D+01	.1000D+01	.1000D+01
5.0	.1000D+01	.1000D+01	.1000D+01	.1000D+01	.1000D+01	.1000D+01
6.0	.1000D+01	.1000D+01	.1000D+01	.1000D+01	.1000D+01	.1000D+01
7.0	.1000D+01	.1000D+01	.1000D+01	.1000D+01	.1000D+01	.1000D+01
8.0	.1000D+01	.1000D+01	.1000D+01	.1000D+01	.1000D+01	.1000D+01
9.0	.1000D+01	.1000D+01	.1000D+01	.1000D+01	.1000D+01	.1000D+01
10.0	.1000D+01	.1000D+01	.1000D+01	.1000D+01	.1000D+01	.1000D+01
15.0	.1000D+01	.1000D+01	.1000D+01	.1000D+01	.1000D+01	.1000D+01
20.0	.1000D+01	.1000D+01	.1000D+01	.1000D+01	.1000D+01	.1000D+01
25.0	.1000D+01	.1000D+01	.1000D+01	.1000D+01	.1000D+01	.1000D+01
30.0	.1000D+01	.1000D+01	.1000D+01	.1000D+01	.1000D+01	.1000D+01
35.0	.1000D+01	.1000D+01	.1000D+01	.1000D+01	.1000D+01	.1000D+01
40.0	.1000D+01	.1000D+01	.1000D+01	.1000D+01	.1000D+01	.1000D+01
45.0	.1000D+01	.1000D+01	.1000D+01	.1000D+01	.1000D+01	.1000D+01
50.0	.1000D+01	.1000D+01	.1000D+01	.1000D+01	.1000D+01	.1000D+01

DIMENSIONLESS CONCENTRATION C/C0 FOR

V = .80 D = 8.00 R = 1.0 LAMBDA = .0 ALPHA = .0 T0 = 60.0

T(YEARS)	X = 400.	X = 500.	X = 600.	X = 700.	X = 800.	X = 1000.
.5	.1019D–05	.2031D–10	.0000D+00	.0000D+00	.0000D+00	.0000D+00
1.0	.7643D–01	.3020D–02	.2471D–04	.3946D–07	.1199D–10	.0000D+00
2.0	.9576D+00	.7842D+00	.4416D+00	.1405D+00	.2236D–01	.5599D–04
3.0	.9999D+00	.9979D+00	.9823D+00	.9102D+00	.7194D+00	.1743D+00
4.0	.1000D+01	.1000D+01	.9999D+00	.9990D+00	.9924D+00	.8662D+00
5.0	.1000D+01	.1000D+01	.1000D+01	.1000D+01	.9999D+00	.9966D+00
6.0	.1000D+01	.1000D+01	.1000D+01	.1000D+01	.1000D+01	.1000D+01
7.0	.1000D+01	.1000D+01	.1000D+01	.1000D+01	.1000D+01	.1000D+01
8.0	.1000D+01	.1000D+01	.1000D+01	.1000D+01	.1000D+01	.1000D+01
9.0	.1000D+01	.1000D+01	.1000D+01	.1000D+01	.1000D+01	.1000D+01
10.0	.1000D+01	.1000D+01	.1000D+01	.1000D+01	.1000D+01	.1000D+01
15.0	.1000D+01	.1000D+01	.1000D+01	.1000D+01	.1000D+01	.1000D+01
20.0	.1000D+01	.1000D+01	.1000D+01	.1000D+01	.1000D+01	.1000D+01
25.0	.1000D+01	.1000D+01	.1000D+01	.1000D+01	.1000D+01	.1000D+01
30.0	.1000D+01	.1000D+01	.1000D+01	.1000D+01	.1000D+01	.1000D+01
35.0	.1000D+01	.1000D+01	.1000D+01	.1000D+01	.1000D+01	.1000D+01
40.0	.1000D+01	.1000D+01	.1000D+01	.1000D+01	.1000D+01	.1000D+01
45.0	.1000D+01	.1000D+01	.1000D+01	.1000D+01	.1000D+01	.1000D+01
50.0	.1000D+01	.1000D+01	.1000D+01	.1000D+01	.1000D+01	.1000D+01

DIMENSIONLESS CONCENTRATION C/C0 FOR

V = .80 D =40.00 R = 1.0 LAMBDA = .0 ALPHA = .0 T0 = 60.0

T(YEARS)	X = 50.	X = 100.	X = 150.	X = 200.	X = 250.	X = 300.
.5	.7892D+00	.6319D+00	.4571D+00	.2944D+00	.1669D+00	.8250D–01
1.0	.9387D+00	.8832D+00	.8048D+00	.7049D+00	.5893D+00	.4672D+00
2.0	.9925D+00	.9848D+00	.9723D+00	.9533D+00	.9261D+00	.8889D+00
3.0	.9989D+00	.9976D+00	.9956D+00	.9922D+00	.9869D+00	.9790D+00
4.0	.9998D+00	.9996D+00	.9992D+00	.9986D+00	.9976D+00	.9961D+00
5.0	.1000D+01	.9999D+00	.9999D+00	.9997D+00	.9996D+00	.9992D+00
6.0	.1000D+01	.1000D+01	.1000D+01	.1000D+01	.9999D+00	.9999D+00
7.0	.1000D+01	.1000D+01	.1000D+01	.1000D+01	.1000D+01	.1000D+01
8.0	.1000D+01	.1000D+01	.1000D+01	.1000D+01	.1000D+01	.1000D+01
9.0	.1000D+01	.1000D+01	.1000D+01	.1000D+01	.1000D+01	.1000D+01
10.0	.1000D+01	.1000D+01	.1000D+01	.1000D+01	.1000D+01	.1000D+01
15.0	.1000D+01	.1000D+01	.1000D+01	.1000D+01	.1000D+01	.1000D+01
20.0	.1000D+01	.1000D+01	.1000D+01	.1000D+01	.1000D+01	.1000D+01
25.0	.1000D+01	.1000D+01	.1000D+01	.1000D+01	.1000D+01	.1000D+01
30.0	.1000D+01	.1000D+01	.1000D+01	.1000D+01	.1000D+01	.1000D+01
35.0	.1000D+01	.1000D+01	.1000D+01	.1000D+01	.1000D+01	.1000D+01
40.0	.1000D+01	.1000D+01	.1000D+01	.1000D+01	.1000D+01	.1000D+01
45.0	.1000D+01	.1000D+01	.1000D+01	.1000D+01	.1000D+01	.1000D+01
50.0	.1000D+01	.1000D+01	.1000D+01	.1000D+01	.1000D+01	.1000D+01

DIMENSIONLESS CONCENTRATION C/C0 FOR

V = .80 D = 40.00 R = 1.0 LAMBDA = .0 ALPHA = .0 T0 = 60.0

T(YEARS)	X = 400.	X = 500.	X = 600.	X = 700.	X = 800.	X = 1000.
.5	.1311D–01	.1135D–02	.5238D–04	.1269D–05	.1602D–07	.3516D–12
1.0	.2455D+00	.9949D–01	.3024D–01	.6809D–02	.1125D–02	.1180D–04
2.0	.7816D+00	.6357D+00	.4679D+00	.3068D+00	.1771D+00	.3881D–01
3.0	.9524D+00	.9048D+00	.8302D+00	.7273D+00	.6013D+00	.3324D+00
4.0	.9902D+00	.9783D+00	.9564D+00	.9198D+00	.8643D+00	.6909D+00
5.0	.9980D+00	.9953D+00	.9897D+00	.9794D+00	.9615D+00	.8901D+00
6.0	.9996D+00	.9990D+00	.9977D+00	.9951D+00	.9901D+00	.9666D+00
7.0	.9999D+00	.9998D+00	.9995D+00	.9989D+00	.9976D+00	.9908D+00
8.0	.1000D+01	.1000D+01	.9999D+00	.9997D+00	.9994D+00	.9976D+00
9.0	.1000D+01	.1000D+01	.1000D+01	.9999D+00	.9999D+00	.9994D+00
10.0	.1000D+01	.1000D+01	.1000D+01	.1000D+01	.1000D+01	.9999D+00
15.0	.1000D+01	.1000D+01	.1000D+01	.1000D+01	.1000D+01	.1000D+01
20.0	.1000D+01	.1000D+01	.1000D+01	.1000D+01	.1000D+01	.1000D+01
25.0	.1000D+01	.1000D+01	.1000D+01	.1000D+01	.1000D+01	.1000D+01
30.0	.1000D+01	.1000D+01	.1000D+01	.1000D+01	.1000D+01	.1000D+01
35.0	.1000D+01	.1000D+01	.1000D+01	.1000D+01	.1000D+01	.1000D+01
40.0	.1000D+01	.1000D+01	.1000D+01	.1000D+01	.1000D+01	.1000D+01
45.0	.1000D+01	.1000D+01	.1000D+01	.1000D+01	.1000D+01	.1000D+01
50.0	.1000D+01	.1000D+01	.1000D+01	.1000D+01	.1000D+01	.1000D+01

DIMENSIONLESS CONCENTRATION C/C0 FOR

V = 1.00 D = 1.00 R = 1.0 LAMBDA = .0 ALPHA = .0 T0 = 60.0

T(YEARS)	X = 50.	X = 100.	X = 150.	X = 200.	X = 250.	X = 300.
.5	.1000D+01	.1000D+01	.9566D+00	.1810D+00	.2059D–03	.3835D–09
1.0	.1000D+01	.1000D+01	.1000D+01	.1000D+01	.1000D+01	.9922D+00
2.0	.1000D+01	.1000D+01	.1000D+01	.1000D+01	.1000D+01	.1000D+01
3.0	.1000D+01	.1000D+01	.1000D+01	.1000D+01	.1000D+01	.1000D+01
4.0	.1000D+01	.1000D+01	.1000D+01	.1000D+01	.1000D+01	.1000D+01
5.0	.1000D+01	.1000D+01	.1000D+01	.1000D+01	.1000D+01	.1000D+01
6.0	.1000D+01	.1000D+01	.1000D+01	.1000D+01	.1000D+01	.1000D+01
7.0	.1000D+01	.1000D+01	.1000D+01	.1000D+01	.1000D+01	.1000D+01
8.0	.1000D+01	.1000D+01	.1000D+01	.1000D+01	.1000D+01	.1000D+01
9.0	.1000D+01	.1000D+01	.1000D+01	.1000D+01	.1000D+01	.1000D+01
10.0	.1000D+01	.1000D+01	.1000D+01	.1000D+01	.1000D+01	.1000D+01
15.0	.1000D+01	.1000D+01	.1000D+01	.1000D+01	.1000D+01	.1000D+01
20.0	.1000D+01	.1000D+01	.1000D+01	.1000D+01	.1000D+01	.1000D+01
25.0	.1000D+01	.1000D+01	.1000D+01	.1000D+01	.1000D+01	.1000D+01
30.0	.1000D+01	.1000D+01	.1000D+01	.1000D+01	.1000D+01	.1000D+01
35.0	.1000D+01	.1000D+01	.1000D+01	.1000D+01	.1000D+01	.1000D+01
40.0	.1000D+01	.1000D+01	.1000D+01	.1000D+01	.1000D+01	.1000D+01
45.0	.1000D+01	.1000D+01	.1000D+01	.1000D+01	.1000D+01	.1000D+01
50.0	.1000D+01	.1000D+01	.1000D+01	.1000D+01	.1000D+01	.1000D+01

DIMENSIONLESS CONCENTRATION C/C0 FOR

V = 1.00 D = 1.00 R = 1.0 LAMBDA = .0 ALPHA = .0 T0 = 60.0

T(YEARS)	X = 400.	X = 500.	X = 600.	X = 700.	X = 800.	X = 1000.
.5	.0000D+00	.0000D+00	.0000D+00	.0000D+00	.0000D+00	.0000D+00
1.0	.9898D–01	.3010D–06	.0000D+00	.0000D+00	.0000D+00	.0000D+00
2.0	.1000D+01	.1000D+01	.9997D+00	.7877D+00	.3442D–01	.8679D–12
3.0	.1000D+01	.1000D+01	.1000D+01	.1000D+01	.1000D+01	.9796D+00
4.0	.1000D+01	.1000D+01	.1000D+01	.1000D+01	.1000D+01	.1000D+01
5.0	.1000D+01	.1000D+01	.1000D+01	.1000D+01	.1000D+01	.1000D+01
6.0	.1000D+01	.1000D+01	.1000D+01	.1000D+01	.1000D+01	.1000D+01
7.0	.1000D+01	.1000D+01	.1000D+01	.1000D+01	.1000D+01	.1000D+01
8.0	.1000D+01	.1000D+01	.1000D+01	.1000D+01	.1000D+01	.1000D+01
9.0	.1000D+01	.1000D+01	.1000D+01	.1000D+01	.1000D+01	.1000D+01
10.0	.1000D+01	.1000D+01	.1000D+01	.1000D+01	.1000D+01	.1000D+01
15.0	.1000D+01	.1000D+01	.1000D+01	.1000D+01	.1000D+01	.1000D+01
20.0	.1000D+01	.1000D+01	.1000D+01	.1000D+01	.1000D+01	.1000D+01
25.0	.1000D+01	.1000D+01	.1000D+01	.1000D+01	.1000D+01	.1000D+01
30.0	.1000D+01	.1000D+01	.1000D+01	.1000D+01	.1000D+01	.1000D+01
35.0	.1000D+01	.1000D+01	.1000D+01	.1000D+01	.1000D+01	.1000D+01
40.0	.1000D+01	.1000D+01	.1000D+01	.1000D+01	.1000D+01	.1000D+01
45.0	.1000D+01	.1000D+01	.1000D+01	.1000D+01	.1000D+01	.1000D+01
50.0	.1000D+01	.1000D+01	.1000D+01	.1000D+01	.1000D+01	.1000D+01

DIMENSIONLESS CONCENTRATION C/C0 FOR

V = 1.00 D = 10.00 R = 1.0 LAMBDA = .0 ALPHA = .0 T0 = 60.0

T(YEARS)	X = 50.	X = 100.	X = 150.	X = 200.	X = 250.	X = 300.
.5	.9894D+00	.9205D+00	.7075D+00	.3817D+00	.1270D+00	.2408D–01
1.0	.9999D+00	.9993D+00	.9950D+00	.9756D+00	.9144D+00	.7798D+00
2.0	.1000D+01	.1000D+01	.1000D+01	.1000D+01	.1000D+01	.9998D+00
3.0	.1000D+01	.1000D+01	.1000D+01	.1000D+01	.1000D+01	.1000D+01
4.0	.1000D+01	.1000D+01	.1000D+01	.1000D+01	.1000D+01	.1000D+01
5.0	.1000D+01	.1000D+01	.1000D+01	.1000D+01	.1000D+01	.1000D+01
6.0	.1000D+01	.1000D+01	.1000D+01	.1000D+01	.1000D+01	.1000D+01
7.0	.1000D+01	.1000D+01	.1000D+01	.1000D+01	.1000D+01	.1000D+01
8.0	.1000D+01	.1000D+01	.1000D+01	.1000D+01	.1000D+01	.1000D+01
9.0	.1000D+01	.1000D+01	.1000D+01	.1000D+01	.1000D+01	.1000D+01
10.0	.1000D+01	.1000D+01	.1000D+01	.1000D+01	.1000D+01	.1000D+01
15.0	.1000D+01	.1000D+01	.1000D+01	.1000D+01	.1000D+01	.1000D+01
20.0	.1000D+01	.1000D+01	.1000D+01	.1000D+01	.1000D+01	.1000D+01
25.0	.1000D+01	.1000D+01	.1000D+01	.1000D+01	.1000D+01	.1000D+01
30.0	.1000D+01	.1000D+01	.1000D+01	.1000D+01	.1000D+01	.1000D+01
35.0	.1000D+01	.1000D+01	.1000D+01	.1000D+01	.1000D+01	.1000D+01
40.0	.1000D+01	.1000D+01	.1000D+01	.1000D+01	.1000D+01	.1000D+01
45.0	.1000D+01	.1000D+01	.1000D+01	.1000D+01	.1000D+01	.1000D+01
50.0	.1000D+01	.1000D+01	.1000D+01	.1000D+01	.1000D+01	.1000D+01

DIMENSIONLESS CONCENTRATION C/C0 FOR

V = 1.00 D = 10.00 R = 1.0 LAMBDA = .0 ALPHA = .0 T0 = 60.0

T(YEARS)	X = 400.	X = 500.	X = 600.	X = 700.	X = 800.	X = 1000.
.5	.1366D–03	.5843D–07	.1765D–11	.0000D+00	.0000D+00	.0000D+00
1.0	.3394D+00	.5551D–01	.2806D–02	.4013D–04	.1556D–06	.4338D–13
2.0	.9971D+00	.9728D+00	.8615D+00	.5998D+00	.2811D+00	.1251D–01
3.0	.1000D+01	.1000D+01	.9996D+00	.9964D+00	.9777D+00	.7419D+00
4.0	.1000D+01	.1000D+01	.1000D+01	.1000D+01	.9999D+00	.9966D+00
5.0	.1000D+01	.1000D+01	.1000D+01	.1000D+01	.1000D+01	.1000D+01
6.0	.1000D+01	.1000D+01	.1000D+01	.1000D+01	.1000D+01	.1000D+01
7.0	.1000D+01	.1000D+01	.1000D+01	.1000D+01	.1000D+01	.1000D+01
8.0	.1000D+01	.1000D+01	.1000D+01	.1000D+01	.1000D+01	.1000D+01
9.0	.1000D+01	.1000D+01	.1000D+01	.1000D+01	.1000D+01	.1000D+01
10.0	.1000D+01	.1000D+01	.1000D+01	.1000D+01	.1000D+01	.1000D+01
15.0	.1000D+01	.1000D+01	.1000D+01	.1000D+01	.1000D+01	.1000D+01
20.0	.1000D+01	.1000D+01	.1000D+01	.1000D+01	.1000D+01	.1000D+01
25.0	.1000D+01	.1000D+01	.1000D+01	.1000D+01	.1000D+01	.1000D+01
30.0	.1000D+01	.1000D+01	.1000D+01	.1000D+01	.1000D+01	.1000D+01
35.0	.1000D+01	.1000D+01	.1000D+01	.1000D+01	.1000D+01	.1000D+01
40.0	.1000D+01	.1000D+01	.1000D+01	.1000D+01	.1000D+01	.1000D+01
45.0	.1000D+01	.1000D+01	.1000D+01	.1000D+01	.1000D+01	.1000D+01
50.0	.1000D+01	.1000D+01	.1000D+01	.1000D+01	.1000D+01	.1000D+01

DIMENSIONLESS CONCENTRATION C/C0 FOR

V = 1.00 D = 50.00 R = 1.0 LAMBDA = .0 ALPHA = .0 T0 = 60.0

T(YEARS)	X = 50.	X = 100.	X = 150.	X = 200.	X = 250.	X = 300.
.5	.8485D+00	.7270D+00	.5789D+00	.4233D+00	.2814D+00	.1686D+00
1.0	.9648D+00	.9315D+00	.8819D+00	.8143D+00	.7291D+00	.6298D+00
2.0	.9971D+00	.9941D+00	.9890D+00	.9810D+00	.9690D+00	.9516D+00
3.0	.9997D+00	.9994D+00	.9988D+00	.9979D+00	.9963D+00	.9939D+00
4.0	.1000D+01	.9999D+00	.9999D+00	.9997D+00	.9996D+00	.9992D+00
5.0	.1000D+01	.1000D+01	.1000D+01	.1000D+01	.9999D+00	.9999D+00
6.0	.1000D+01	.1000D+01	.1000D+01	.1000D+01	.1000D+01	.1000D+01
7.0	.1000D+01	.1000D+01	.1000D+01	.1000D+01	.1000D+01	.1000D+01
8.0	.1000D+01	.1000D+01	.1000D+01	.1000D+01	.1000D+01	.1000D+01
9.0	.1000D+01	.1000D+01	.1000D+01	.1000D+01	.1000D+01	.1000D+01
10.0	.1000D+01	.1000D+01	.1000D+01	.1000D+01	.1000D+01	.1000D+01
15.0	.1000D+01	.1000D+01	.1000D+01	.1000D+01	.1000D+01	.1000D+01
20.0	.1000D+01	.1000D+01	.1000D+01	.1000D+01	.1000D+01	.1000D+01
25.0	.1000D+01	.1000D+01	.1000D+01	.1000D+01	.1000D+01	.1000D+01
30.0	.1000D+01	.1000D+01	.1000D+01	.1000D+01	.1000D+01	.1000D+01
35.0	.1000D+01	.1000D+01	.1000D+01	.1000D+01	.1000D+01	.1000D+01
40.0	.1000D+01	.1000D+01	.1000D+01	.1000D+01	.1000D+01	.1000D+01
45.0	.1000D+01	.1000D+01	.1000D+01	.1000D+01	.1000D+01	.1000D+01
50.0	.1000D+01	.1000D+01	.1000D+01	.1000D+01	.1000D+01	.1000D+01

DIMENSIONLESS CONCENTRATION C/C0 FOR

V = 1.00 D = 50.00 R = 1.0 LAMBDA = .0 ALPHA = .0 T0 = 60.0

T(YEARS)	X = 400.	X = 500.	X = 600.	X = 700.	X = 800.	X = 1000.
.5	.4352D-01	.7006D-02	.6889D-03	.4082D-04	.1442D-05	.3736D-09
1.0	.4143D+00	.2257D+00	.9915D-01	.3464D-01	.9525D-02	.3413D-03
2.0	.8962D+00	.8086D+00	.6869D+00	.5419D+00	.3921D+00	.1522D+00
3.0	.9854D+00	.9684D+00	.9380D+00	.8893D+00	.8185D+00	.6138D+00
4.0	.9980D+00	.9953D+00	.9897D+00	.9794D+00	.9615D+00	.8901D+00
5.0	.9997D+00	.9993D+00	.9984D+00	.9966D+00	.9930D+00	.9756D+00
6.0	.1000D+01	.9999D+00	.9998D+00	.9995D+00	.9988D+00	.9953D+00
7.0	.1000D+01	.1000D+01	.1000D+01	.9999D+00	.9998D+00	.9991D+00
8.0	.1000D+01	.1000D+01	.1000D+01	.1000D+01	.1000D+01	.9999D+00
9.0	.1000D+01	.1000D+01	.1000D+01	.1000D+01	.1000D+01	.1000D+01
10.0	.1000D+01	.1000D+01	.1000D+01	.1000D+01	.1000D+01	.1000D+01
15.0	.1000D+01	.1000D+01	.1000D+01	.1000D+01	.1000D+01	.1000D+01
20.0	.1000D+01	.1000D+01	.1000D+01	.1000D+01	.1000D+01	.1000D+01
25.0	.1000D+01	.1000D+01	.1000D+01	.1000D+01	.1000D+01	.1000D+01
30.0	.1000D+01	.1000D+01	.1000D+01	.1000D+01	.1000D+01	.1000D+01
35.0	.1000D+01	.1000D+01	.1000D+01	.1000D+01	.1000D+01	.1000D+01
40.0	.1000D+01	.1000D+01	.1000D+01	.1000D+01	.1000D+01	.1000D+01
45.0	.1000D+01	.1000D+01	.1000D+01	.1000D+01	.1000D+01	.1000D+01
50.0	.1000D+01	.1000D+01	.1000D+01	.1000D+01	.1000D+01	.1000D+01

Appendix B

ODAST: A Computer Program for Evaluation of the Analytical Solution for One-Dimensional Contaminant Transport

The program ODAST evaluates the one dimensional analytical solute transport solution given in section 2.2 considering convection, dispersion, decay, and adsorption in porous media. It includes two simple function type subroutines. One subroutine calculates the product of the exponential exp(A) and the complementary error function erfc(B); it was written by *Van Genuchten and Alves* [1982]. The input data for this program is very short and simple. A user's guide for the program and a listing of the code are given below.

User's Guide

The following describes the necessary sequence of cards for running the program ODAST.

	Variable	*Description*
Card 1 (2I5)		
Col. 1-5	NUMX	Number of x positions
6-10	NUMT	Number of time points
Card 2 (6D10.4)		
One card for each six values of x position		
Col. 1-10	X(1)	Values of distance x from the source, m
11-20	X(2)	
21-30	X(3)	
31-40	X(4)	
41-50	X(5)	
51-60	X(6)	
Card 3 (6D10.4)		
One card for each six values of time		
Col. 1-10	T(1)	Values of time at which concentration C/C_o is required, years
11-20	T(2)	
21-30	T(3)	
31-40	T(4)	
41-50	T(5)	
51-60	T(6)	

Card 4 (6D10.4)

Col.		
1-10	D1	Longitudinal dispersion coefficient, m^2/d
11-20	V1	Pore water velocity, m/d
21-30	R	Retardation factor
31-40	T0	Total period of waste recharge, years
41-50	ALAM1	Radioactive decay factor, d^{-1}
51-60	ALFA1	Decay factor of the source, d^{-1}

To solve a series of problems for the same space and time positions, but different values of D1, V1, R, T0, ALAM1, and ALFA1, Card 4 may be repeated as often as necessary. A blank card should follow the last data card, before the end-of-file card.

```
      PROGRAM ODAST
C
C     THIS PROGRAM EVALUATES THE ONE DIMENSIONAL ANALYTICAL
C     SOLUTE TRANSPORT SOLUTION CONSIDERING CONVECTION,
C     DISPERSION, DECAY AND ADSORPTION.
C
C     NUMX   =   NUMBER OF X POSITIONS
C     NUMT   =   NUMBER OF TIMES
C     X      =   DISTANCES FROM THE SOURCE (METERS)
C     T      =   TIME ELAPSED SINCE THE BEGINNING OF
C                  OPERATION (YEARS)
C     D1     =   COEFFICIENT OF DISPERSION (M*M/DAY)
C     V1     =   PORE WATER VELOCITY (M/DAY)
C     R      =   RETARDATION FACTOR
C     T0     =   THE TOTAL PERIOD OF WASTE RECHARGE (YEARS)
C     ALAM1  =   DECAY FACTOR OF THE SOLUTE (1/DAY)
C     ALFA1  =   DECAY FACTOR OF THE SOURCE (1/DAY)
C
      DOUBLE PRECISION X,T,CD,PI,D1,V1,R,T0,ALAM1,ALFA1
     1 ,V,D,ALAM,ALFA,T1,X1,DSQRT2,U,C1,C2,B1,B2,B3,C3,
     2 A1,A,BB1,BB2,DD2,CC1,CC2,DD,DD1,A2,ARG
      DIMENSION CD(6),X(100),T(100)
      DATA IXE,MINUS/2HX=,1H-/
      PI=4*DATAN(1.D0)
      READ(5,510) NUMX,NUMT
      READ(5,520) (X(I),I=1,NUMX)
      READ(5,520) (T(I),I=1,NUMT)
   10 READ(5,520) D1,V1,R,T0,ALAM1,ALFA1
      IF (D1 .EQ. 0) STOP
      V=V1*365.25
      D=D1*365.25
      ALAM=ALAM1*365.25
      ALFA=ALFA1*365.25
      DO 90 IX=1,NUMX,6
        LX=MIN0(IX+5,NUMX)
        IP=MAX0(2-IX,0)
        WRITE(6,610) IP,V1,D1,R,ALAM1,ALFA1,T0,(IXE,X(I),I=IX,LX)
        WRITE(6,620) (MINUS,I=IX,LX)
        DO 80 II=1,NUMT
          T1=T(II)
          DSQRT2=2*DSQRT(D*R*T1)
```

```
          K1=0
          DO 70 KK=IX,LX
             K1=K1+1
             X1=X(KK)
             IFLAG=0
 20          IF (ALFA.EQ.ALAM) GO TO 30
             U=DSQRT(V*V+4*D*R*(ALAM-ALFA))
             C1=X1*(V-U)/(2*D)
             C2=X1*(V+U)/(2*D)
             B1=(R*X1-U*T1)/DSQRT2
             B2=(R*X1+U*T1)/DSQRT2
             B3=(R*X1+V*T1)/DSQRT2
             C3=(V*X1/D)+(ALFA-ALAM)*T1
             A1=(V/(V+U))*EXER(C1,B1)+(V/(V-U))*EXER(C2,B2)+
     1            (V*V/(2*D*R*(ALAM-ALFA)))*EXER(C3,B3)
             A=A1
             GO TO 40
 30          BB1=(R*X1-V*T1)/DSQRT2
             BB2=V*X1/D
             DD2=DSQRT(V*V*T1/(PI*D*R))
             CC1=BB1*BB1
             CC2=(R*X1+V*T1)/DSQRT2
             DD=1.+(V*X1/D)+(V*V*T1/(D*R))
             DD1=0.0
             A2=0.5*EXER(DD1,BB1)+DD2*EXPD(-CC1)-0.5*DD*EXER(BB2,CC2)
             A=A2
 40          IF (IFLAG.EQ.1) GO TO 50
             ARG=-T1*ALFA
             CD(K1)=EXPD(ARG)*A
             IF (T1.LE.T0) GO TO 60
             T1=T1-T0
             IFLAG=1
             GO TO 20
 50          CD(K1)=CD(K1)-A*EXPD(-ALFA*T(II))
 60          IF (CD(K1).LT.1.E-14) CD(K1)=0.
 70          CONTINUE
          WRITE(6,630) T(II),(CD(KK),KK=1,K1)
 80       CONTINUE
 90     CONTINUE
      GO TO 10
510   FORMAT(2I5)
520   FORMAT(6D10.4)
610   FORMAT(//I1,20X,36HDIMENSIONLESS CONCENTRATION C/C0 FOR
     1 /1H0,7X,2HV=,F6.2,2X,2HD=,F6.2,2X,2HR=,F3.1,2X,7HLAMBDA=
     2 ,F5.3,2X,6HALPHA=,F5.3,
     3 2X,3HT0=,F5.1/8X,61(1H-)/,9H T(YEARS),6(4X,A2,F5.0))
620   FORMAT(1X,10(1H-),6(A1,10(1H-)))
630   FORMAT(3X,F5.1,2X,6D11.4)
      END
```

```
      DOUBLE PRECISION FUNCTION EXER(A,B)
C
C     THIS SUBROUTINE IS FROM VAN GENUCHTEN AND ALVES (1982)
C
C     PURPOSE.. TO CALCULATE EXP(A)*ERFC(B)
C
      DOUBLE PRECISION A,B,C,X,T,Y
      EXER=0.0
      IF ((DABS(A).GT.170.).AND.B.LE.0.) RETURN
      IF (B.NE.0.0) GO TO 100
      EXER=EXPD(A)
      RETURN
  100 C=A-B*B
      IF ((DABS(C).GT.170.).AND.(B.GT.0)) RETURN
      IF (C.LT.-170.) GO TO 130
      X=DABS(B)
      IF (X.GT.3.0) GO TO 110
      T=1./(1.+.3275911*X)
      Y=T*(.2548296-T*(.2844967-T*(1.421414-T*(1.453152-1.061405*T))))
      GO TO 120
  110 Y=.5641896/(X+.5/(X+1./(X+1.5/(X+2./(X+2.5/(X+1.))))))
  120 EXER=Y*EXPD(C)
  130 IF (B.LT.0.0) EXER=2*EXPD(A)-EXER
      RETURN
      END
```

```
      DOUBLE PRECISION FUNCTION EXPD(X)
C
C     THIS IS THE DOUBLE PRECISION EXPONENTIAL FUNCTION
C     EXCEPT THAT ARGUMENTS LESS THAN -170 IMPLY A VALUE
C     OF 0.  THIS IS TO PREVENT AN ERROR MESSAGE IN DEXP.
C
      DOUBLE PRECISION X
      EXPD=0.0
      IF (X.LT.-170) RETURN
      EXPD=DEXP(X)
      RETURN
      END
```

Appendix C

Tables of Dimensionless Concentration for Two–Dimensional Contaminant Transport in Aquifers With One–Dimensional Uniform Flow

The following tables list dimensionless concentration C/C_o (evaluated from (48)) for the following values: time t = 100–1825 days; velocity v = 0.1, 0.5, 1 m/d; half length of source a = 50 m; longitudinal dispersion coefficient D_L = 1.0–50 m^2/d; transverse dispersion coefficient D_T = 0.1–5 m^2/d; retardation factor R = 1; decay factor of source $\alpha = 0$ d^{-1}; decay factor of solute $\lambda = 0$ d^{-1}; and C/C_o is given as a function of distance downstream from the source, x in meters for different values of distance from the x axis, y in meters.

VALUES OF CONCENTATION(C/C0) AT TIME T= 100.0 DAYS
****

V= .100 A= 50.0 DL= 1.00 DT= .10 R= 1.00
ALFA= .0000 LAMBDA= .0000

X	Y = .0	Y = 5.0	Y = 10.0	Y = 20.0	Y = 30.0	Y = 40.0
10.0	.71379	.71379	.71379	.71379	.71379	.71270
15.0	.53461	.53461	.53461	.53461	.53461	.53322
20.0	.36498	.36498	.36498	.36498	.36498	.36361
25.0	.22561	.22561	.22561	.22561	.22561	.22451
30.0	.12563	.12563	.12563	.12563	.12563	.12489
35.0	.06277	.06277	.06277	.06277	.06277	.06277
40.0	.02806	.02806	.02806	.02806	.02806	.02784
45.0	.01119	.01119	.01119	.01119	.01119	.01110
50.0	.00398	.00398	.00398	.00398	.00398	.00394
60.0	.00035	.00035	.00035	.00035	.00035	.00035
75.0	.00000	.00000	.00000	.00000	.00000	.00000
100.0	.00000	.00000	.00000	.00000	.00000	.00000
125.0	.00000	.00000	.00000	.00000	.00000	.00000
150.0	.00000	.00000	.00000	.00000	.00000	.00000
175.0	.00000	.00000	.00000	.00000	.00000	.00000
200.0	.00000	.00000	.00000	.00000	.00000	.00000

X	Y = 45.0	Y = 50.0	Y = 55.0	Y = 60.0	Y = 70.0	Y = 80.0
10.0	.68470	.35687	.02904	.00102	-.00014	-.00026
15.0	.50210	.26730	.03251	.00139	.00000	-.00001
20.0	.33698	.18249	.02800	.00137	.00000	.00000
25.0	.20561	.11280	.02000	.00110	.00000	.00000
30.0	.11337	.06281	.01225	.00074	.00000	.00000
35.0	.05623	.03138	.00654	.00043	.00000	.00000
40.0	.02499	.01403	.00306	.00021	.00000	.00000
45.0	.00993	.00560	.00126	.00009	.00000	.00000
50.0	.00352	.00199	.00046	.00003	.00000	.00000
60.0	.00031	.00018	.00004	.00000	.00000	.00000
75.0	.00000	.00000	.00000	.00000	.00000	.00000
100.0	.00000	.00000	.00000	.00000	.00000	.00000
125.0	.00000	.00000	.00000	.00000	.00000	.00000
150.0	.00000	.00000	.00000	.00000	.00000	.00000
175.0	.00000	.00000	.00000	.00000	.00000	.00000
200.0	.00000	.00000	.00000	.00000	.00000	.00000

VALUES OF CONCENTATION(C/C0) AT TIME T= 365.0 DAYS
****

V = .100 A = 50.0 DL = 1.00 DT = .10 R = 1.00

ALFA= .0000 LAMBDA= .0000

X	Y = .0	Y = 5.0	Y = 10.0	Y = 20.0	Y = 30.0	Y = 40.0
10.0	.95155	.95155	.95155	.95154	.95125	.93901
15.0	.91385	.91385	.91385	.91384	.91334	.89358
20.0	.86421	.86421	.86421	.86420	.86345	.83670
25.0	.80389	.80389	.80389	.80388	.80291	.77060
30.0	.73409	.73409	.73409	.73408	.73290	.69705
35.0	.65687	.65687	.65687	.65685	.65552	.61831
40.0	.57498	.57498	.57498	.57497	.57356	.53701
45.0	.49164	.49164	.49164	.49163	.49021	.45601
50.0	.41011	.41011	.41011	.41009	.40873	.37808
60.0	.26377	.26377	.26377	.26376	.26267	.24079
75.0	.11034	.11034	.11034	.11033	.10976	.09968
100.0	.01419	.01419	.01419	.01419	.01410	.01269
125.0	.00083	.00083	.00083	.00083	.00083	.00074
150.0	.00002	.00002	.00002	.00002	.00002	.00002
175.0	.00000	.00000	.00000	.00000	.00000	.00000
200.0	.00000	.00000	.00000	.00000	.00000	.00000

X	Y = 45.0	Y = 50.0	Y = 55.0	Y = 60.0	Y = 70.0	Y = 80.0
10.0	.87593	.47574	.07554	.01244	.00013	-.00029
15.0	.80644	.45692	.10741	.02027	.00051	.00000
20.0	.73576	.43210	.12845	.02751	.00075	.00001
25.0	.66482	.40195	.13907	.03329	.00099	.00001
30.0	.59320	.36705	.14089	.03704	.00119	.00001
35.0	.52109	.32843	.13578	.03856	.00134	.00002
40.0	.44944	.28749	.12554	.03797	.00143	.00002
45.0	.37975	.24582	.11190	.03563	.00144	.00002
50.0	.31371	.20505	.09639	.03202	.00138	.00002
60.0	.19887	.13189	.06490	.02298	.00110	.00002
75.0	.08204	.05517	.02830	.01066	.00058	.00001
100.0	.01042	.00710	.00377	.00150	.00009	.00000
125.0	.00061	.00042	.00022	.00009	.00001	.00000
150.0	.00002	.00001	.00001	.00000	.00000	.00000
175.0	.00000	.00000	.00000	.00000	.00000	.00000
200.0	.00000	.00000	.00000	.00000	.00000	.00000

VALUES OF CONCENTATION(C/C0) AT TIME T= 730.0 DAYS
****

V = .100 A = 50.0 DL = 1.00 DT = .10 R = 1.00

ALFA = .0000 LAMBDA = .0000

X	Y = .0	Y = 5.0	Y = 10.0	Y = 20.0	Y = 30.0	Y = 40.0
10.0	.98999	.98999	.98999	.98993	.98883	.97152
15.0	.98318	.98318	.98318	.98306	.98111	.95219
20.0	.97248	.97248	.97247	.97229	.96928	.92818
25.0	.95837	.95837	.95836	.95810	.95385	.90104
30.0	.94034	.94034	.94032	.93997	.93437	.87109
35.0	.91793	.91793	.91791	.91746	.91046	.83841
40.0	.89082	.89081	.89079	.89022	.88185	.80302
45.0	.85879	.85879	.85876	.85808	.84843	.76488
50.0	.82183	.82183	.82180	.82101	.81024	.72401
60.0	.73401	.73400	.73396	.73299	.72070	.63471
75.0	.57618	.57617	.57613	.57504	.56254	.48785
100.0	.30559	.30559	.30555	.30470	.29613	.25305
125.0	.11624	.11624	.11622	.11581	.11206	.09503
150.0	.03067	.03067	.03067	.03054	.02947	.02488
175.0	.00550	.00550	.00550	.00548	.00527	.00444
200.0	.00066	.00066	.00066	.00066	.00063	.00053

X	Y = 45.0	Y = 50.0	Y = 55.0	Y = 60.0	Y = 70.0	Y = 80.0
10.0	.90265	.49496	.08726	.01838	.00099	-.00022
15.0	.85462	.49159	.12856	.03099	.00207	.00011
20.0	.81098	.48624	.16150	.04430	.00320	.00019
25.0	.77208	.47918	.18628	.05732	.00452	.00027
30.0	.73631	.47017	.20403	.06925	.00597	.00037
35.0	.70210	.45897	.21583	.07952	.00747	.00048
40.0	.66823	.44541	.22259	.08780	.00897	.00059
45.0	.63382	.42940	.22497	.09391	.01036	.00071
50.0	.59832	.41092	.22351	.09782	.01159	.00082
60.0	.52307	.36700	.21094	.09930	.01330	.00101
75.0	.40171	.28809	.17447	.08833	.01364	.00114
100.0	.20863	.15280	.09697	.05255	.00946	.00090
125.0	.07847	.05812	.03777	.02121	.00418	.00043
150.0	.02057	.01534	.01010	.00579	.00121	.00013
175.0	.00368	.00275	.00183	.00106	.00023	.00003
200.0	.00044	.00033	.00022	.00013	.00003	.00000

VALUES OF CONCENTATION(C/C0) AT TIME T= 1825.0 DAYS

V = .100 A = 50.0 DL = 1.00 DT = .10 R = 1.00

ALFA = .0000 LAMBDA = .0000

X	Y = .0	Y = 25.0	Y = 40.0	Y = 50.0	Y = 60.0	Y = 70.0
25.0	.99853	.99590	.93172	.49928	.06684	.00757
50.0	.99454	.98473	.85602	.49735	.13867	.02495
75.0	.98028	.95965	.79510	.49036	.18561	.04653
100.0	.94528	.91354	.73570	.47309	.21048	.06539
125.0	.87603	.83639	.66261	.43874	.21487	.07670
150.0	.76362	.72168	.56636	.38276	.19916	.07787
175.0	.61249	.57425	.44821	.30727	.16633	.06912
200.0	.44322	.41306	.32137	.22253	.12369	.05358
225.0	.28473	.26418	.20514	.14306	.08097	.03610
250.0	.16037	.14830	.11502	.08062	.04622	.02103
275.0	.07846	.07237	.05609	.03946	.02283	.01055
300.0	.03311	.03048	.02361	.01666	.00971	.00453
325.0	.01199	.01103	.00854	.00604	.00353	.00166
350.0	.00371	.00341	.00264	.00187	.00110	.00052
375.0	.00098	.00090	.00070	.00049	.00029	.00014
400.0	.00022	.00020	.00016	.00011	.00007	.00003

X	Y = 80.0	Y = 90.0	Y = 100.0	Y = 110.0	Y = 120.0	Y = 130.0
25.0	.00095	.00012	.00002	.00000	.00000	.00000
50.0	.00389	.00056	.00007	.00001	.00000	.00000
75.0	.00909	.00150	.00021	.00003	.00000	.00000
100.0	.01533	.00289	.00045	.00006	.00001	.00000
125.0	.02061	.00433	.00073	.00010	.00001	.00000
150.0	.02308	.00529	.00095	.00014	.00002	.00000
175.0	.02196	.00538	.00102	.00015	.00002	.00000
200.0	.01789	.00460	.00092	.00014	.00002	.00000
225.0	.01249	.00334	.00069	.00011	.00001	.00000
250.0	.00747	.00205	.00044	.00007	.00001	.00000
275.0	.00382	.00107	.00023	.00004	.00001	.00000
300.0	.00166	.00048	.00011	.00002	.00000	.00000
325.0	.00062	.00018	.00004	.00001	.00000	.00000
350.0	.00020	.00006	.00001	.00000	.00000	.00000
375.0	.00005	.00002	.00000	.00000	.00000	.00000
400.0	.00001	.00000	.00000	.00000	.00000	.00000

VALUES OF CONCENTATION(C/C0) AT TIME T= 100.0 DAYS
****

V = .100 A = 50.0 DL = 5.00 DT = .50 R = 1.00

ALFA = .0000 LAMBDA = .0000

X	Y = .0	Y = 25.0	Y = 40.0	Y = 50.0	Y = 60.0	Y = 70.0
25.0	.53887	.53845	.50426	.26944	.03461	.00216
50.0	.18148	.18114	.16195	.09074	.01954	.00164
75.0	.03603	.03592	.03139	.01801	.00464	.00047
100.0	.00408	.00406	.00351	.00204	.00057	.00006
125.0	.00026	.00026	.00022	.00013	.00004	.00000
150.0	.00001	.00001	.00001	.00000	.00000	.00000
175.0	.00000	.00000	.00000	.00000	.00000	.00000
200.0	.00000	.00000	.00000	.00000	.00000	.00000
225.0	.00000	.00000	.00000	.00000	.00000	.00000
250.0	.00000	.00000	.00000	.00000	.00000	.00000
275.0	.00000	.00000	.00000	.00000	.00000	.00000
300.0	.00000	.00000	.00000	.00000	.00000	.00000
325.0	.00000	.00000	.00000	.00000	.00000	.00000
350.0	.00000	.00000	.00000	.00000	.00000	.00000
375.0	.00000	.00000	.00000	.00000	.00000	.00000
400.0	.00000	.00000	.00000	.00000	.00000	.00000

X	Y = 80.0	Y = 90.0	Y = 100.0	Y = 110.0	Y = 120.0	Y = 130.0
25.0	.00007	.00000	.00000	.00000	.00000	.00000
50.0	.00006	.00000	.00000	.00000	.00000	.00000
75.0	.00002	.00000	.00000	.00000	.00000	.00000
100.0	.00000	.00000	.00000	.00000	.00000	.00000
125.0	.00000	.00000	.00000	.00000	.00000	.00000
150.0	.00000	.00000	.00000	.00000	.00000	.00000
175.0	.00000	.00000	.00000	.00000	.00000	.00000
200.0	.00000	.00000	.00000	.00000	.00000	.00000
225.0	.00000	.00000	.00000	.00000	.00000	.00000
250.0	.00000	.00000	.00000	.00000	.00000	.00000
275.0	.00000	.00000	.00000	.00000	.00000	.00000
300.0	.00000	.00000	.00000	.00000	.00000	.00000
325.0	.00000	.00000	.00000	.00000	.00000	.00000
350.0	.00000	.00000	.00000	.00000	.00000	.00000
375.0	.00000	.00000	.00000	.00000	.00000	.00000
400.0	.00000	.00000	.00000	.00000	.00000	.00000

VALUES OF CONCENTATION(C/C0) AT TIME T= 365.0 DAYS
****

V = .100 A = 50.0 DL = 5.00 DT = .50 R = 1.00
ALFA = .0000 LAMBDA = .0000

X	Y = .0	Y = 25.0	Y = 40.0	Y = 50.0	Y = 60.0	Y = 70.0
25.0	.82962	.81892	.72943	.41498	.10053	.02310
50.0	.61782	.60015	.49659	.30924	.12188	.03601
75.0	.40679	.38942	.31157	.20377	.09597	.03356
100.0	.23415	.22159	.17456	.11738	.06021	.02335
125.0	.11685	.10964	.08572	.05862	.03153	.01304
150.0	.05024	.04685	.03648	.02523	.01397	.00603
175.0	.01853	.01720	.01337	.00931	.00525	.00233
200.0	.00584	.00541	.00419	.00294	.00168	.00076
225.0	.00157	.00145	.00112	.00079	.00046	.00021
250.0	.00036	.00033	.00026	.00018	.00011	.00005
275.0	.00007	.00006	.00005	.00004	.00002	.00001
300.0	.00001	.00001	.00001	.00001	.00000	.00000
325.0	.00000	.00000	.00000	.00000	.00000	.00000
350.0	.00000	.00000	.00000	.00000	.00000	.00000
375.0	.00000	.00000	.00000	.00000	.00000	.00000
400.0	.00000	.00000	.00000	.00000	.00000	.00000

X	Y = 80.0	Y = 90.0	Y = 100.0	Y = 110.0	Y = 120.0	Y = 130.0
25.0	.00517	.00103	.00017	.00002	.00000	.00000
50.0	.00894	.00187	.00032	.00005	.00001	.00000
75.0	.00925	.00206	.00037	.00005	.00001	.00000
100.0	.00700	.00165	.00031	.00005	.00001	.00000
125.0	.00416	.00103	.00020	.00003	.00000	.00000
150.0	.00201	.00052	.00010	.00002	.00000	.00000
175.0	.00081	.00022	.00004	.00001	.00000	.00000
200.0	.00027	.00007	.00002	.00000	.00000	.00000
225.0	.00008	.00002	.00000	.00000	.00000	.00000
250.0	.00002	.00001	.00000	.00000	.00000	.00000
275.0	.00000	.00000	.00000	.00000	.00000	.00000
300.0	.00000	.00000	.00000	.00000	.00000	.00000
325.0	.00000	.00000	.00000	.00000	.00000	.00000
350.0	.00000	.00000	.00000	.00000	.00000	.00000
375.0	.00000	.00000	.00000	.00000	.00000	.00000
400.0	.00000	.00000	.00000	.00000	.00000	.00000

VALUES OF CONCENTATION(C/C0) AT TIME T= 730.0 DAYS
****

V = .100 A = 50.0 DL = 5.00 DT = .50 R = 1.00
ALFA = .0000 LAMBDA = .0000

X	Y = .0	Y = 25.0	Y = 40.0	Y = 50.0	Y = 60.0	Y = 70.0
25.0	.91719	.89706	.79008	.46005	.13001	.03985
50.0	.80374	.76596	.62532	.40497	.18461	.07175
75.0	.66835	.62250	.49258	.33859	.18458	.08429
100.0	.52443	.48001	.37532	.26722	.15909	.08032
125.0	.38654	.34942	.27209	.19807	.12403	.06677
150.0	.26660	.23893	.18586	.13733	.08877	.04988
175.0	.17153	.15282	.11890	.08877	.05862	.03394
200.0	.10269	.09111	.07094	.05336	.03577	.02116
225.0	.05708	.05050	.03935	.02977	.02018	.01212
250.0	.02941	.02596	.02025	.01538	.01051	.00639
275.0	.01403	.01236	.00965	.00735	.00506	.00310
300.0	.00619	.00545	.00425	.00325	.00225	.00139
325.0	.00252	.00222	.00173	.00133	.00092	.00057
350.0	.00095	.00083	.00065	.00050	.00035	.00022
375.0	.00033	.00029	.00023	.00017	.00012	.00008
400.0	.00010	.00009	.00007	.00006	.00004	.00002

X	Y = 80.0	Y = 90.0	Y = 100.0	Y = 110.0	Y = 120.0	Y = 130.0
25.0	.01338	.00451	.00146	.00044	.00012	.00003
50.0	.02654	.00937	.00311	.00095	.00027	.00007
75.0	.03440	.01286	.00442	.00139	.00039	.00010
100.0	.03549	.01403	.00500	.00161	.00047	.00012
125.0	.03133	.01299	.00481	.00159	.00047	.00012
150.0	.02446	.01055	.00403	.00137	.00041	.00011
175.0	.01720	.00766	.00301	.00104	.00032	.00009
200.0	.01099	.00502	.00202	.00072	.00022	.00006
225.0	.00642	.00299	.00123	.00044	.00014	.00004
250.0	.00344	.00163	.00068	.00025	.00008	.00002
275.0	.00169	.00081	.00034	.00013	.00004	.00001
300.0	.00076	.00037	.00016	.00006	.00002	.00001
325.0	.00032	.00015	.00007	.00003	.00001	.00000
350.0	.00012	.00006	.00003	.00001	.00000	.00000
375.0	.00004	.00002	.00001	.00000	.00000	.00000
400.0	.00001	.00001	.00000	.00000	.00000	.00000

VALUES OF CONCENTATION(C/C0) AT TIME T= 1825.0 DAYS
****

V = .100 A = 50.0 DL = 5.00 DT = .50 R = 1.00

ALFA = .0000 LAMBDA = .0000

X	Y = .0	Y = 25.0	Y = 40.0	Y = 50.0	Y = 60.0	Y = 70.0
25.0	.96996	.94349	.82751	.49029	.15299	.05621
50.0	.92746	.87484	.71312	.47596	.23860	.11024
75.0	.87329	.80292	.63816	.45640	.27431	.14840
100.0	.80919	.73076	.57786	.43135	.28436	.17008
125.0	.73731	.65844	.52202	.40096	.27930	.17843
150.0	.66001	.58571	.46678	.36586	.26422	.17662
175.0	.57975	.51289	.41114	.32709	.24226	.16726
200.0	.49904	.44099	.35551	.28606	.21582	.15261
225.0	.42037	.37147	.30103	.24438	.18696	.13463
250.0	.34606	.30600	.24912	.20368	.15751	.11503
275.0	.27806	.24610	.20117	.16542	.12902	.09528
300.0	.21781	.19299	.15831	.13079	.10271	.07653
325.0	.16617	.14739	.12128	.10058	.07944	.05962
350.0	.12335	.10953	.09036	.07518	.05966	.04505
375.0	.08902	.07913	.06543	.05459	.04348	.03300
400.0	.06242	.05554	.04602	.03848	.03075	.02344

X	Y = 80.0	Y = 90.0	Y = 100.0	Y = 110.0	Y = 120.0	Y = 130.0
25.0	.02429	.01131	.00544	.00263	.00127	.00060
50.0	.05223	.02543	.01253	.00616	.00299	.00143
75.0	.07730	.03976	.02026	.01019	.00502	.00242
100.0	.09578	.05201	.02750	.01418	.00712	.00348
125.0	.10669	.06075	.03328	.01762	.00903	.00447
150.0	.11049	.06545	.03703	.02011	.01051	.00529
175.0	.10829	.06625	.03854	.02142	.01141	.00583
200.0	.10145	.06370	.03795	.02154	.01168	.00605
225.0	.09136	.05858	.03562	.02059	.01134	.00596
250.0	.07934	.05176	.03200	.01880	.01051	.00559
275.0	.06658	.04405	.02763	.01645	.00931	.00502
300.0	.05405	.03618	.02297	.01384	.00792	.00431
325.0	.04248	.02871	.01841	.01121	.00648	.00356
350.0	.03232	.02202	.01425	.00875	.00510	.00283
375.0	.02382	.01634	.01065	.00659	.00387	.00216
400.0	.01700	.01173	.00770	.00479	.00284	.00159

VALUES OF CONCENTATION(C/C0) AT TIME T= 100.0 DAYS
****

V = .500 A = 50.0 DL = 5.00 DT = .50 R = 1.00

ALFA = .0000 LAMBDA = .0000

X	Y = .0	Y = 20.0	Y = 40.0	Y = 60.0	Y = 80.0	Y = 100.0
20.0	.92779	.92774	.89147	.03631	.00005	.00000
40.0	.74493	.74480	.68332	.06161	.00013	.00000
60.0	.47762	.47746	.42465	.05297	.00016	.00000
80.0	.23012	.23000	.20082	.02930	.00012	.00000
100.0	.08007	.08001	.06910	.01097	.00005	.00000
120.0	.01963	.01961	.01682	.00281	.00002	.00000
140.0	.00334	.00334	.00285	.00049	.00000	.00000
160.0	.00039	.00039	.00033	.00006	.00000	.00000
180.0	.00003	.00003	.00003	.00000	.00000	.00000
200.0	.00000	.00000	.00000	.00000	.00000	.00000
220.0	.00000	.00000	.00000	.00000	.00000	.00000
240.0	.00000	.00000	.00000	.00000	.00000	.00000
260.0	.00000	.00000	.00000	.00000	.00000	.00000
280.0	.00000	.00000	.00000	.00000	.00000	.00000
300.0	.00000	.00000	.00000	.00000	.00000	.00000
320.0	.00000	.00000	.00000	.00000	.00000	.00000

X	Y = 120.0	Y = 140.0	Y = 160.0	Y = 180.0	Y = 200.0	Y = 220.0
20.0	-.00001	-.00003	-.00007	-.00015	-.00030	-.00053
40.0	.00000	.00000	.00000	.00000	.00000	.00000
60.0	.00000	.00000	.00000	.00000	.00000	.00000
80.0	.00000	.00000	.00000	.00000	.00000	.00000
100.0	.00000	.00000	.00000	.00000	.00000	.00000
120.0	.00000	.00000	.00000	.00000	.00000	.00000
140.0	.00000	.00000	.00000	.00000	.00000	.00000
160.0	.00000	.00000	.00000	.00000	.00000	.00000
180.0	.00000	.00000	.00000	.00000	.00000	.00000
200.0	.00000	.00000	.00000	.00000	.00000	.00000
220.0	.00000	.00000	.00000	.00000	.00000	.00000
240.0	.00000	.00000	.00000	.00000	.00000	.00000
260.0	.00000	.00000	.00000	.00000	.00000	.00000
280.0	.00000	.00000	.00000	.00000	.00000	.00000
300.0	.00000	.00000	.00000	.00000	.00000	.00000
320.0	.00000	.00000	.00000	.00000	.00000	.00000

VALUES OF CONCENTATION(C/C0) AT TIME T= 365.0 DAYS
****

V = .500 A = 50.0 DL = 5.00 DT = .50 R = 1.00
ALFA = .0000 LAMBDA = .0000

X	Y = .0	Y = 20.0	Y = 40.0	Y = 60.0	Y = 80.0	Y = 100.0
20.0	.99933	.99871	.94876	.05059	.00063	.00001
40.0	.99704	.99470	.88432	.11281	.00243	.00004
60.0	.99051	.98501	.83040	.16034	.00574	.00012
80.0	.97537	.96557	.78363	.19225	.01031	.00025
100.0	.94528	.93084	.73570	.21048	.01533	.00045
120.0	.89319	.87479	.67891	.21563	.01973	.00067
140.0	.81386	.79308	.60785	.20776	.02251	.00088
160.0	.70699	.68591	.52123	.18776	.02306	.00100
180.0	.57920	.55987	.42294	.15828	.02134	.00102
200.0	.44322	.42715	.32137	.12369	.01789	.00092
220.0	.31426	.30214	.22666	.08908	.01359	.00074
240.0	.20511	.19682	.14735	.05883	.00935	.00053
260.0	.12259	.11746	.08780	.03548	.00582	.00035
280.0	.06682	.06394	.04774	.01948	.00328	.00020
300.0	.03311	.03165	.02361	.00971	.00166	.00011
320.0	.01488	.01421	.01059	.00438	.00076	.00005

X	Y = 120.0	Y = 140.0	Y = 160.0	Y = 180.0	Y = 200.0	Y = 220.0
20.0	-.00001	-.00003	-.00007	-.00015	-.00030	-.00053
40.0	.00000	.00000	.00000	.00000	.00000	.00000
60.0	.00000	.00000	.00000	.00000	.00000	.00000
80.0	.00000	.00000	.00000	.00000	.00000	.00000
100.0	.00001	.00000	.00000	.00000	.00000	.00000
120.0	.00001	.00000	.00000	.00000	.00000	.00000
140.0	.00001	.00000	.00000	.00000	.00000	.00000
160.0	.00002	.00000	.00000	.00000	.00000	.00000
180.0	.00002	.00000	.00000	.00000	.00000	.00000
200.0	.00002	.00000	.00000	.00000	.00000	.00000
220.0	.00001	.00000	.00000	.00000	.00000	.00000
240.0	.00001	.00000	.00000	.00000	.00000	.00000
260.0	.00001	.00000	.00000	.00000	.00000	.00000
280.0	.00000	.00000	.00000	.00000	.00000	.00000
300.0	.00000	.00000	.00000	.00000	.00000	.00000
320.0	.00000	.00000	.00000	.00000	.00000	.00000

VALUES OF CONCENTATION(C/C0) AT TIME T= 730.0 DAYS
****

V = .500 A = 50.0 DL = 5.00 DT = .50 R = 1.00

ALFA = .0000 LAMBDA = .0000

X	Y = .0	Y = 20.0	Y = 40.0	Y = 60.0	Y = 80.0	Y = 100.0
20.0	.99992	.99927	.94917	.05078	.00068	.00001
40.0	.99985	.99735	.88628	.11370	.00264	.00007
60.0	.99956	.99353	.83672	.16323	.00642	.00019
80.0	.99894	.98775	.80007	.19979	.01209	.00046
100.0	.99779	.98024	.77231	.22731	.01934	.00092
120.0	.99581	.97130	.75042	.24863	.02768	.00162
140.0	.99262	.96109	.73233	.26548	.03661	.00260
160.0	.98763	.94946	.71650	.27884	.04567	.00386
180.0	.97996	.93587	.70154	.28912	.05447	.00536
200.0	.96836	.91931	.68602	.29637	.06262	.00703
220.0	.95116	.89828	.66839	.30026	.06974	.00876
240.0	.92637	.87097	.64692	.30025	.07539	.01043
260.0	.89186	.83539	.61991	.29565	.07918	.01188
280.0	.84582	.78982	.58587	.28583	.08073	.01298
300.0	.78720	.73322	.54389	.27040	.07979	.01359
320.0	.71619	.66572	.49394	.24943	.07630	.01364

X	Y = 120.0	Y = 140.0	Y = 160.0	Y = 180.0	Y = 200.0	Y = 220.0
20.0	-.00001	-.00003	-.00007	-.00015	-.00030	-.00053
40.0	.00000	.00000	.00000	.00000	.00000	.00000
60.0	.00001	.00000	.00000	.00000	.00000	.00000
80.0	.00002	.00000	.00000	.00000	.00000	.00000
100.0	.00003	.00000	.00000	.00000	.00000	.00000
120.0	.00007	.00000	.00000	.00000	.00000	.00000
140.0	.00013	.00000	.00000	.00000	.00000	.00000
160.0	.00021	.00001	.00000	.00000	.00000	.00000
180.0	.00033	.00001	.00000	.00000	.00000	.00000
200.0	.00049	.00002	.00000	.00000	.00000	.00000
220.0	.00067	.00003	.00000	.00000	.00000	.00000
240.0	.00086	.00005	.00000	.00000	.00000	.00000
260.0	.00106	.00006	.00000	.00000	.00000	.00000
280.0	.00123	.00007	.00000	.00000	.00000	.00000
300.0	.00137	.00008	.00000	.00000	.00000	.00000
320.0	.00144	.00009	.00000	.00000	.00000	.00000

VALUES OF CONCENTATION(C/C0) AT TIME T= 1825.0 DAYS
****

V = .500 A = 50.0 DL = 5.00 DT = .50 R = 1.00

ALFA = .0000 LAMBDA = .0000

X	Y = .0	Y = 20.0	Y = 40.0	Y = 60.0	Y = 80.0	Y = 100.0
20.0	.99993	.99928	.94918	.05078	.00068	.00001
40.0	.99987	.99736	.88629	.11371	.00264	.00007
60.0	.99961	.99357	.83675	.16325	.00642	.00020
80.0	.99907	.98787	.80016	.19984	.01211	.00046
100.0	.99814	.98056	.77255	.22745	.01940	.00093
120.0	.99668	.97210	.75102	.24897	.02782	.00166
140.0	.99462	.96293	.73372	.26627	.03692	.00269
160.0	.99191	.95341	.71947	.28052	.04635	.00404
180.0	.98854	.94377	.70748	.29250	.05583	.00573
200.0	.98454	.93419	.69722	.30274	.06520	.00773
220.0	.97995	.92477	.68831	.31163	.07433	.01003
240.0	.97483	.91556	.68047	.31942	.08318	.01258
260.0	.96926	.90660	.67351	.32634	.09168	.01537
280.0	.96330	.89790	.66726	.33251	.09984	.01835
300.0	.95702	.88946	.66161	.33808	.10764	.02149
320.0	.95047	.88126	.65645	.34312	.11510	.02476

X	Y = 120.0	Y = 140.0	Y = 160.0	Y = 180.0	Y = 200.0	Y = 220.0
20.0	−.00001	−.00003	−.00007	−.00015	−.00030	−.00053
40.0	.00000	.00000	.00000	.00000	.00000	.00000
60.0	.00001	.00000	.00000	.00000	.00000	.00000
80.0	.00002	.00000	.00000	.00000	.00000	.00000
100.0	.00004	.00000	.00000	.00000	.00000	.00000
120.0	.00008	.00000	.00000	.00000	.00000	.00000
140.0	.00014	.00001	.00000	.00000	.00000	.00000
160.0	.00025	.00001	.00000	.00000	.00000	.00000
180.0	.00040	.00002	.00000	.00000	.00000	.00000
200.0	.00061	.00004	.00000	.00000	.00000	.00000
220.0	.00090	.00006	.00000	.00000	.00000	.00000
240.0	.00126	.00010	.00001	.00000	.00000	.00000
260.0	.00172	.00014	.00001	.00000	.00000	.00000
280.0	.00226	.00021	.00002	.00000	.00000	.00000
300.0	.00290	.00029	.00002	.00000	.00000	.00000
320.0	.00363	.00039	.00003	.00000	.00000	.00000

VALUES OF CONCENTATION(C/C0) AT TIME T= 100.0 DAYS
****

V = .500 A = 50.0 DL = 25.00 DT = 2.50 R = 1.00

ALFA = .0000 LAMBDA = .0000

X	Y = .0	Y = 20.0	Y = 40.0	Y = 60.0	Y = 80.0	Y = 100.0
20.0	.90561	.89967	.80900	.09746	.00677	.00042
40.0	.78046	.76878	.63625	.14597	.01341	.00088
60.0	.64015	.62496	.49323	.14941	.01763	.00124
80.0	.49629	.48054	.36943	.12968	.01852	.00141
100.0	.36224	.34829	.26383	.10115	.01663	.00137
120.0	.24811	.23720	.17813	.07231	.01319	.00116
140.0	.15906	.15138	.11309	.04773	.00941	.00088
160.0	.09524	.09032	.06725	.02919	.00610	.00060
180.0	.05316	.05028	.03736	.01655	.00362	.00038
200.0	.02763	.02607	.01935	.00871	.00197	.00021
220.0	.01336	.01258	.00933	.00425	.00099	.00011
240.0	.00600	.00564	.00418	.00192	.00046	.00005
260.0	.00250	.00235	.00174	.00081	.00020	.00002
280.0	.00097	.00091	.00067	.00031	.00008	.00001
300.0	.00035	.00033	.00024	.00011	.00003	.00000
320.0	.00012	.00011	.00008	.00004	.00001	.00000

X	Y = 120.0	Y = 140.0	Y = 160.0	Y = 180.0	Y = 200.0	Y = 220.0
20.0	.00002	.00000	−.00002	−.00005	−.00013	−.00027
40.0	.00003	.00000	.00000	.00000	.00000	.00000
60.0	.00005	.00000	.00000	.00000	.00000	.00000
80.0	.00006	.00000	.00000	.00000	.00000	.00000
100.0	.00006	.00000	.00000	.00000	.00000	.00000
120.0	.00005	.00000	.00000	.00000	.00000	.00000
140.0	.00004	.00000	.00000	.00000	.00000	.00000
160.0	.00003	.00000	.00000	.00000	.00000	.00000
180.0	.00002	.00000	.00000	.00000	.00000	.00000
200.0	.00001	.00000	.00000	.00000	.00000	.00000
220.0	.00001	.00000	.00000	.00000	.00000	.00000
240.0	.00000	.00000	.00000	.00000	.00000	.00000
260.0	.00000	.00000	.00000	.00000	.00000	.00000
280.0	.00000	.00000	.00000	.00000	.00000	.00000
300.0	.00000	.00000	.00000	.00000	.00000	.00000
320.0	.00000	.00000	.00000	.00000	.00000	.00000

VALUES OF CONCENTATION(C/C0) AT TIME T= 365.0 DAYS
****

V = .500 A = 50.0 DL = 25.00 DT = 2.50 R = 1.00

ALFA = .0000 LAMBDA = .0000

X	Y = .0	Y = 20.0	Y = 40.0	Y = 60.0	Y = 80.0	Y = 100.0
20.0	.97877	.96728	.86018	.12673	.01890	.00419
40.0	.94593	.92168	.75206	.21238	.04108	.00955
60.0	.90711	.87164	.68024	.25708	.06288	.01561
80.0	.86120	.81771	.62539	.27791	.08158	.02178
100.0	.80919	.76123	.57786	.28436	.09578	.02750
120.0	.75219	.70291	.53307	.28124	.10511	.03228
140.0	.69143	.64321	.48891	.27123	.10976	.03579
160.0	.62812	.58265	.44457	.25612	.11026	.03790
180.0	.56356	.52188	.39998	.23726	.10725	.03858
200.0	.49904	.46174	.35551	.21582	.10145	.03795
220.0	.43583	.40312	.31176	.19284	.09357	.03620
240.0	.37513	.34699	.26948	.16926	.08430	.03357
260.0	.31801	.29423	.22939	.14592	.07426	.03032
280.0	.26535	.24561	.19216	.12355	.06403	.02671
300.0	.21781	.20170	.15831	.10271	.05405	.02297
320.0	.17579	.16287	.12820	.08382	.04469	.01930

X	Y = 120.0	Y = 140.0	Y = 160.0	Y = 180.0	Y = 200.0	Y = 220.0
20.0	.00097	.00021	.00003	-.00005	-.00013	-.00027
40.0	.00226	.00050	.00010	.00002	.00000	.00000
60.0	.00378	.00084	.00017	.00003	.00000	.00000
80.0	.00545	.00123	.00025	.00004	.00001	.00000
100.0	.00712	.00164	.00033	.00006	.00001	.00000
120.0	.00867	.00204	.00042	.00007	.00001	.00000
140.0	.00998	.00240	.00050	.00009	.00001	.00000
160.0	.01094	.00270	.00057	.00010	.00001	.00000
180.0	.01151	.00290	.00062	.00011	.00002	.00000
200.0	.01168	.00301	.00065	.00012	.00002	.00000
220.0	.01145	.00301	.00066	.00012	.00002	.00000
240.0	.01089	.00292	.00065	.00012	.00002	.00000
260.0	.01006	.00275	.00062	.00012	.00002	.00000
280.0	.00905	.00251	.00057	.00011	.00002	.00000
300.0	.00792	.00223	.00052	.00010	.00002	.00000
320.0	.00677	.00194	.00045	.00009	.00001	.00000

VALUES OF CONCENTATION(C/C0) AT TIME T= 730.0 DAYS
****

V = .500 A = 50.0 DL = 25.00 DT = 2.50 R = 1.00
ALFA = .0000 LAMBDA = .0000

X	Y = .0	Y = 20.0	Y = 40.0	Y = 60.0	Y = 80.0	Y = 100.0
20.0	.98701	.97505	.86668	.13156	.02206	.00601
40.0	.96554	.94017	.76755	.22387	.04860	.01388
60.0	.94154	.90411	.70744	.27726	.07609	.02322
80.0	.91402	.86752	.66714	.30890	.10188	.03349
100.0	.88388	.83169	.63692	.32822	.12454	.04411
120.0	.85192	.79699	.61196	.33986	.14358	.05453
140.0	.81874	.76334	.58969	.34617	.15900	.06432
160.0	.78476	.73046	.56864	.34847	.17102	.07317
180.0	.75021	.69805	.54793	.34750	.17990	.08084
200.0	.71522	.66582	.52701	.34376	.18591	.08720
220.0	.67987	.63355	.50554	.33758	.18930	.09218
240.0	.64417	.60109	.48333	.32921	.19032	.09574
260.0	.60815	.56835	.46028	.31889	.18917	.09790
280.0	.57188	.53528	.43638	.30680	.18606	.09873
300.0	.53542	.50194	.41169	.29316	.18121	.09828
320.0	.49889	.46840	.38632	.27820	.17483	.09667

X	Y = 120.0	Y = 140.0	Y = 160.0	Y = 180.0	Y = 200.0	Y = 220.0
20.0	.00190	.00062	.00019	.00001	-.00011	-.00026
40.0	.00445	.00148	.00049	.00016	.00005	.00001
60.0	.00765	.00258	.00086	.00028	.00008	.00002
80.0	.01140	.00391	.00132	.00043	.00013	.00004
100.0	.01558	.00545	.00186	.00061	.00019	.00005
120.0	.02002	.00717	.00248	.00082	.00025	.00007
140.0	.02456	.00900	.00316	.00105	.00033	.00010
160.0	.02901	.01090	.00388	.00131	.00041	.00012
180.0	.03323	.01279	.00463	.00158	.00050	.00015
200.0	.03706	.01461	.00538	.00185	.00059	.00018
220.0	.04040	.01630	.00610	.00213	.00069	.00021
240.0	.04316	.01780	.00678	.00239	.00078	.00023
260.0	.04529	.01906	.00738	.00263	.00087	.00026
280.0	.04674	.02006	.00788	.00285	.00095	.00029
300.0	.04752	.02077	.00828	.00302	.00101	.00031
320.0	.04763	.02117	.00856	.00316	.00107	.00033

VALUES OF CONCENTATION(C/C0) AT TIME T= 1825.0 DAYS
****

V = .500 A = 50.0 DL = 25.00 DT = 2.50 R = 1.00
ALFA = .0000 LAMBDA = .0000

X	Y = .0	Y = 20.0	Y = 40.0	Y = 60.0	Y = 80.0	Y = .100.0
20.0	.98844	.97642	.86792	.13259	.02286	.00659
40.0	.96899	.94350	.77054	.22636	.05053	.01528
60.0	.94772	.91007	.71280	.28173	.07957	.02574
80.0	.92378	.87694	.67559	.31595	.10737	.03747
100.0	.89821	.84552	.64933	.33859	.13261	.04995
120.0	.87195	.81631	.62930	.35435	.15486	.06271
140.0	.84571	.78935	.61305	.36570	.17421	.07536
160.0	.82003	.76449	.59920	.37402	.19092	.08763
180.0	.79523	.74149	.58695	.38015	.20534	.09934
200.0	.77151	.72014	.57582	.38461	.21777	.11038
220.0	.74896	.70023	.56548	.38777	.22848	.12070
240.0	.72759	.68161	.55573	.38987	.23770	.13027
260.0	.70735	.66410	.54641	.39109	.24562	.13910
280.0	.68820	.64759	.53743	.39158	.25241	.14721
300.0	.67006	.63194	.52872	.39142	.25819	.15461
320.0	.65284	.61707	.52021	.39069	.26306	.16133

X	Y = 120.0	Y = 140.0	Y = 160.0	Y = 180.0	Y = 200.0	Y = 220.0
20.0	.00229	.00087	.00034	.00009	-.00007	-.00024
40.0	.00540	.00208	.00084	.00035	.00015	.00007
60.0	.00935	.00365	.00150	.00063	.00027	.00012
80.0	.01409	.00561	.00232	.00099	.00043	.00019
100.0	.01953	.00795	.00334	.00143	.00062	.00027
120.0	.02556	.01067	.00455	.00198	.00087	.00038
140.0	.03203	.01373	.00597	.00262	.00116	.00052
160.0	.03881	.01711	.00758	.00338	.00151	.00068
180.0	.04578	.02076	.00938	.00424	.00192	.00087
200.0	.05280	.02462	.01136	.00522	.00239	.00109
220.0	.05980	.02865	.01349	.00629	.00291	.00134
240.0	.06668	.03280	.01577	.00747	.00350	.00163
260.0	.07339	.03703	.01817	.00875	.00415	.00195
280.0	.07987	.04128	.02066	.01011	.00486	.00230
300.0	.08607	.04552	.02322	.01154	.00562	.00269
320.0	.09198	.04970	.02583	.01304	.00643	.00311

VALUES OF CONCENTATION(C/C0) AT TIME T= 100.0 DAYS
****

V = 1.000 A = 50.0 DL = 10.00 DT = 1.00 R = 1.00
ALFA = .0000 LAMBDA = .0000

X	Y = .0	Y = 20.0	Y = 40.0	Y = 60.0	Y = 80.0	Y = 100.0
20.0	.99030	.98993	.94213	.04817	.00037	.00000
40.0	.95777	.95648	.85525	.10253	.00130	.00001
60.0	.88435	.88174	.75191	.13246	.00263	.00001
80.0	.75755	.75385	.62285	.13473	.00374	.00002
100.0	.58524	.58125	.47056	.11473	.00404	.00002
120.0	.39798	.39457	.31526	.08276	.00345	.00002
140.0	.23377	.23143	.18333	.05048	.00237	.00002
160.0	.11697	.11567	.09110	.02590	.00132	.00001
180.0	.04937	.04877	.03826	.01112	.00061	.00001
200.0	.01745	.01723	.01347	.00398	.00023	.00000
220.0	.00514	.00507	.00396	.00118	.00007	.00000
240.0	.00126	.00124	.00097	.00029	.00002	.00000
260.0	.00025	.00025	.00020	.00006	.00000	.00000
280.0	.00004	.00004	.00003	.00001	.00000	.00000
300.0	.00001	.00001	.00000	.00000	.00000	.00000
320.0	.00000	.00000	.00000	.00000	.00000	.00000

X	Y = 120.0	Y = 140.0	Y = 160.0	Y = 180.0	Y = 200.0	Y = 220.0
20.0	-.00001	-.00002	-.00006	-.00013	-.00027	-.00048
40.0	.00000	.00000	.00000	.00000	.00000	.00000
60.0	.00000	.00000	.00000	.00000	.00000	.00000
80.0	.00000	.00000	.00000	.00000	.00000	.00000
100.0	.00000	.00000	.00000	.00000	.00000	.00000
120.0	.00000	.00000	.00000	.00000	.00000	.00000
140.0	.00000	.00000	.00000	.00000	.00000	.00000
160.0	.00000	.00000	.00000	.00000	.00000	.00000
180.0	.00000	.00000	.00000	.00000	.00000	.00000
200.0	.00000	.00000	.00000	.00000	.00000	.00000
220.0	.00000	.00000	.00000	.00000	.00000	.00000
240.0	.00000	.00000	.00000	.00000	.00000	.00000
260.0	.00000	.00000	.00000	.00000	.00000	.00000
280.0	.00000	.00000	.00000	.00000	.00000	.00000
300.0	.00000	.00000	.00000	.00000	.00000	.00000
320.0	.00000	.00000	.00000	.00000	.00000	.00000

VALUES OF CONCENTATION(C/C0) AT TIME T= 365.0 DAYS
****

V = 1.000 A = 50.0 DL = 10.00 DT = 1.00 R = 1.00
ALFA = .0000 LAMBDA = .0000

X	Y = .0	Y = 20.0	Y = 40.0	Y = 60.0	Y = 80.0	Y = 100.0
20.0	1.00015	.99950	.94940	.05078	.00068	.00001
40.0	.99985	.99735	.88628	.11370	.00264	.00007
60.0	.99956	.99353	.83672	.16323	.00642	.00019
80.0	.99894	.98775	.80007	.19979	.01209	.00046
100.0	.99779	.98024	.77231	.22731	.01934	.00092
120.0	.99581	.97130	.75042	.24863	.02768	.00162
140.0	.99262	.96109	.73233	.26548	.03661	.00260
160.0	.98763	.94946	.71650	.27884	.04567	.00386
180.0	.97996	.93587	.70154	.28912	.05447	.00536
200.0	.96836	.91931	.68602	.29637	.06262	.00703
220.0	.95116	.89828	.66839	.30026	.06974	.00876
240.0	.92637	.87097	.64692	.30025	.07539	.01043
260.0	.89186	.83539	.61991	.29565	.07918	.01188
280.0	.84582	.78982	.58587	.28583	.08073	.01298
300.0	.78720	.73322	.54389	.27040	.07979	.01359
320.0	.71619	.66572	.49394	.24943	.07630	.01364

X	Y = 120.0	Y = 140.0	Y = 160.0	Y = 180.0	Y = 200.0	Y = 220.0
20.0	-.00001	-.00002	-.00006	-.00013	-.00027	-.00048
40.0	.00000	.00000	.00000	.00000	.00000	.00000
60.0	.00001	.00000	.00000	.00000	.00000	.00000
80.0	.00002	.00000	.00000	.00000	.00000	.00000
100.0	.00003	.00000	.00000	.00000	.00000	.00000
120.0	.00007	.00000	.00000	.00000	.00000	.00000
140.0	.00013	.00000	.00000	.00000	.00000	.00000
160.0	.00021	.00001	.00000	.00000	.00000	.00000
180.0	.00033	.00001	.00000	.00000	.00000	.00000
200.0	.00049	.00002	.00000	.00000	.00000	.00000
220.0	.00067	.00003	.00000	.00000	.00000	.00000
240.0	.00086	.00005	.00000	.00000	.00000	.00000
260.0	.00106	.00006	.00000	.00000	.00000	.00000
280.0	.00123	.00007	.00000	.00000	.00000	.00000
300.0	.00137	.00008	.00000	.00000	.00000	.00000
320.0	.00144	.00009	.00000	.00000	.00000	.00000

VALUES OF CONCENTATION(C/C0) AT TIME T= 730.0 DAYS
****

V = 1.000 A = 50.0 DL = 10.00 DT = 1.00 R = 1.00

ALFA = .0000 LAMBDA = .0000

X	Y = .0	Y = 20.0	Y = 40.0	Y = 60.0	Y = 80.0	Y = 100.0
20.0	1.00015	.99950	.94940	.05078	.00068	.00001
40.0	.99986	.99736	.88629	.11371	.00264	.00007
60.0	.99961	.99357	.83675	.16325	.00642	.00020
80.0	.99907	.98787	.80016	.19984	.01211	.00046
100.0	.99814	.98056	.77255	.22745	.01940	.00093
120.0	.99668	.97210	.75102	.24897	.02782	.00166
140.0	.99462	.96293	.73372	.26627	.03692	.00269
160.0	.99191	.95341	.71947	.28052	.04635	.00404
180.0	.98854	.94377	.70748	.29250	.05583	.00573
200.0	.98454	.93419	.69721	.30274	.06520	.00773
220.0	.97994	.92476	.68830	.31163	.07433	.01003
240.0	.97482	.91555	.68047	.31942	.08317	.01258
260.0	.96924	.90658	.67350	.32633	.09168	.01537
280.0	.96326	.89786	.66723	.33249	.09983	.01834
300.0	.95693	.88938	.66155	.33804	.10762	.02148
320.0	.95031	.88112	.65633	.34304	.11506	.02474

X	Y = 120.0	Y = 140.0	Y = 160.0	Y = 180.0	Y = 200.0	Y = 220.0
20.0	-.00001	-.00002	-.00006	-.00013	-.00027	-.00048
40.0	.00000	.00000	.00000	.00000	.00000	.00000
60.0	.00001	.00000	.00000	.00000	.00000	.00000
80.0	.00002	.00000	.00000	.00000	.00000	.00000
100.0	.00004	.00000	.00000	.00000	.00000	.00000
120.0	.00008	.00000	.00000	.00000	.00000	.00000
140.0	.00014	.00001	.00000	.00000	.00000	.00000
160.0	.00025	.00001	.00000	.00000	.00000	.00000
180.0	.00040	.00002	.00000	.00000	.00000	.00000
200.0	.00061	.00004	.00000	.00000	.00000	.00000
220.0	.00090	.00006	.00000	.00000	.00000	.00000
240.0	.00126	.00010	.00001	.00000	.00000	.00000
260.0	.00172	.00014	.00001	.00000	.00000	.00000
280.0	.00226	.00021	.00002	.00000	.00000	.00000
300.0	.00289	.00029	.00002	.00000	.00000	.00000
320.0	.00362	.00039	.00003	.00000	.00000	.00000

VALUES OF CONCENTATION(C/C0) AT TIME T= 1825.0 DAYS
****

V = 1.000 A = 50.0 DL = 10.00 DT = 1.00 R = 1.00
ALFA = .0000 LAMBDA = .0000

X	Y = .0	Y = 20.0	Y = 40.0	Y = 60.0	Y = 80.0	Y = 100.0
20.0	1.00015	.99950	.94940	.05078	.00068	.00001
40.0	.99986	.99736	.88629	.11371	.00264	.00007
60.0	.99961	.99357	.83675	.16325	.00642	.00020
80.0	.99907	.98787	.80016	.19984	.01211	.00046
100.0	.99814	.98056	.77255	.22745	.01940	.00093
120.0	.99668	.97210	.75102	.24897	.02782	.00166
140.0	.99462	.96293	.73372	.26627	.03692	.00269
160.0	.99191	.95341	.71947	.28052	.04635	.00404
180.0	.98854	.94377	.70748	.29250	.05583	.00573
200.0	.98454	.93419	.69722	.30274	.06520	.00773
220.0	.97995	.92477	.68831	.31163	.07433	.01003
240.0	.97483	.91556	.68047	.31942	.08318	.01258
260.0	.96926	.90660	.67351	.32634	.09168	.01537
280.0	.96330	.89790	.66726	.33251	.09984	.01835
300.0	.95702	.88946	.66161	.33808	.10764	.02149
320.0	.95047	.88126	.65645	.34312	.11510	.02476

X	Y = 120.0	Y = 140.0	Y = 160.0	Y = 180.0	Y = 200.0	Y = 220.0
20.0	-.00001	-.00002	-.00006	-.00013	-.00027	-.00048
40.0	.00000	.00000	.00000	.00000	.00000	.00000
60.0	.00001	.00000	.00000	.00000	.00000	.00000
80.0	.00002	.00000	.00000	.00000	.00000	.00000
100.0	.00004	.00000	.00000	.00000	.00000	.00000
120.0	.00008	.00000	.00000	.00000	.00000	.00000
140.0	.00014	.00001	.00000	.00000	.00000	.00000
160.0	.00025	.00001	.00000	.00000	.00000	.00000
180.0	.00040	.00002	.00000	.00000	.00000	.00000
200.0	.00061	.00004	.00000	.00000	.00000	.00000
220.0	.00090	.00006	.00000	.00000	.00000	.00000
240.0	.00126	.00010	.00001	.00000	.00000	.00000
260.0	.00172	.00014	.00001	.00000	.00000	.00000
280.0	.00226	.00021	.00002	.00000	.00000	.00000
300.0	.00290	.00029	.00002	.00000	.00000	.00000
320.0	.00363	.00039	.00003	.00000	.00000	.00000

VALUES OF CONCENTATION(C/C0) AT TIME T= 100.0 DAYS
****

V = 1.000 A = 50.0 DL = 50.00 DT = 5.00 R = 1.00

ALFA = .0000 LAMBDA = .0000

X	Y = .0	Y = 20.0	Y = 40.0	Y = 60.0	Y = 80.0	Y = 100.0
20.0	.95919	.94929	.84646	.11675	.01373	.00205
40.0	.89598	.87546	.71580	.18921	.02900	.00455
60.0	.82312	.79396	.61944	.21778	.04238	.00711
80.0	.73860	.70432	.53662	.22046	.05158	.00931
100.0	.64642	.61068	.45994	.20797	.05582	.01084
120.0	.55094	.51674	.38717	.18661	.05549	.01153
140.0	.45662	.42597	.31856	.16059	.05160	.01138
160.0	.36753	.34151	.25536	.13303	.04535	.01054
180.0	.28694	.26587	.19893	.10623	.03793	.00921
200.0	.21706	.20070	.15034	.08184	.03031	.00765
220.0	.15895	.14675	.11006	.06082	.02320	.00604
240.0	.11257	.10382	.07797	.04361	.01703	.00456
260.0	.07705	.07101	.05339	.03016	.01201	.00329
280.0	.05094	.04692	.03532	.02011	.00814	.00228
300.0	.03251	.02993	.02255	.01293	.00530	.00151
320.0	.02001	.01842	.01389	.00801	.00332	.00096

X	Y = 120.0	Y = 140.0	Y = 160.0	Y = 180.0	Y = 200.0	Y = 220.0
20.0	.00027	.00001	-.00003	-.00004	-.00006	-.00009
40.0	.00061	.00006	.00001	.00000	.00000	.00000
60.0	.00098	.00011	.00001	.00000	.00000	.00000
80.0	.00133	.00014	.00001	.00000	.00000	.00000
100.0	.00160	.00018	.00001	.00000	.00000	.00000
120.0	.00177	.00020	.00002	.00000	.00000	.00000
140.0	.00181	.00021	.00002	.00000	.00000	.00000
160.0	.00174	.00021	.00002	.00000	.00000	.00000
180.0	.00157	.00019	.00002	.00000	.00000	.00000
200.0	.00134	.00017	.00001	.00000	.00000	.00000
220.0	.00109	.00014	.00001	.00000	.00000	.00000
240.0	.00084	.00011	.00001	.00000	.00000	.00000
260.0	.00062	.00008	.00001	.00000	.00000	.00000
280.0	.00044	.00006	.00001	.00000	.00000	.00000
300.0	.00030	.00004	.00000	.00000	.00000	.00000
320.0	.00019	.00003	.00000	.00000	.00000	.00000

VALUES OF CONCENTATION(C/C0) AT TIME T= 365.0 DAYS
****

V = 1.000 A = 50.0 DL = 50.00 DT = 5.00 R = 1.00
ALFA = .0000 LAMBDA = .0000

X	Y = .0	Y = 20.0	Y = 40.0	Y = 60.0	Y = 80.0	Y = 100.0
20.0	.98868	.97672	.86835	.13151	.02206	.00601
40.0	.96514	.93978	.76714	.22388	.04860	.01388
60.0	.94150	.90408	.70741	.27726	.07609	.02322
80.0	.91402	.86753	.66714	.30889	.10188	.03349
100.0	.88388	.83169	.63692	.32822	.12454	.04411
120.0	.85192	.79699	.61196	.33986	.14358	.05453
140.0	.81874	.76334	.58969	.34617	.15900	.06432
160.0	.78476	.73046	.56864	.34847	.17102	.07317
180.0	.75021	.69805	.54793	.34750	.17990	.08084
200.0	.71522	.66582	.52701	.34376	.18591	.08720
220.0	.67987	.63355	.50555	.33758	.18930	.09218
240.0	.64417	.60109	.48333	.32921	.19032	.09574
260.0	.60815	.56835	.46028	.31889	.18917	.09790
280.0	.57188	.53528	.43638	.30680	.18606	.09873
300.0	.53542	.50194	.41169	.29316	.18121	.09828
320.0	.49889	.46840	.38632	.27820	.17483	.09667

X	Y = 120.0	Y = 140.0	Y = 160.0	Y = 180.0	Y = 200.0	Y = 220.0
20.0	.00189	.00061	.00018	.00002	-.00004	-.00008
40.0	.00445	.00148	.00049	.00016	.00005	.00001
60.0	.00765	.00258	.00086	.00028	.00008	.00002
80.0	.01140	.00391	.00132	.00043	.00013	.00004
100.0	.01558	.00545	.00186	.00061	.00019	.00005
120.0	.02002	.00717	.00248	.00082	.00025	.00007
140.0	.02456	.00900	.00316	.00105	.00033	.00010
160.0	.02901	.01090	.00388	.00131	.00041	.00012
180.0	.03323	.01279	.00463	.00158	.00050	.00015
200.0	.03706	.01461	.00538	.00185	.00059	.00018
220.0	.04040	.01630	.00610	.00213	.00069	.00021
240.0	.04316	.01780	.00678	.00239	.00078	.00023
260.0	.04529	.01906	.00738	.00263	.00087	.00026
280.0	.04674	.02006	.00788	.00285	.00095	.00029
300.0	.04752	.02077	.00828	.00302	.00101	.00031
320.0	.04763	.02117	.00856	.00316	.00107	.00033

VALUES OF CONCENTATION(C/C0) AT TIME T= 730.0 DAYS
****

V = 1.000 A = 50.0 DL = 50.00 DT = 5.00 R = 1.00

ALFA = .0000 LAMBDA = .0000

X	Y = .0	Y = 20.0	Y = 40.0	Y = 60.0	Y = 80.0	Y = 100.0
20.0	.99006	.97804	.86954	.13250	.02282	.00656
40.0	.96845	.94297	.77000	.22626	.05044	.01521
60.0	.94744	.90980	.71254	.28153	.07940	.02560
80.0	.92339	.87656	.67523	.31563	.10710	.03725
100.0	.89762	.84494	.64879	.33811	.13219	.04962
120.0	.87110	.81549	.62854	.35367	.15427	.06223
140.0	.84455	.78822	.61199	.36475	.17340	.07470
160.0	.81847	.76296	.59777	.37275	.18984	.08674
180.0	.79318	.73948	.58508	.37847	.20391	.09817
200.0	.76886	.71754	.57339	.38244	.21592	.10887
220.0	.74558	.69692	.56239	.38500	.22611	.11877
240.0	.72333	.67744	.55183	.38639	.23473	.12785
260.0	.70206	.65893	.54157	.38677	.24193	.13609
280.0	.68170	.64122	.53149	.38626	.24787	.14350
300.0	.66214	.62420	.52148	.38495	.25266	.15009
320.0	.64329	.60773	.51148	.38289	.25640	.15588

X	Y = 120.0	Y = 140.0	Y = 160.0	Y = 180.0	Y = 200.0	Y = 220.0
20.0	.00226	.00084	.00031	.00009	-.00001	-.00006
40.0	.00534	.00204	.00081	.00033	.00014	.00006
60.0	.00924	.00357	.00144	.00059	.00025	.00010
80.0	.01391	.00548	.00223	.00092	.00039	.00016
100.0	.01927	.00776	.00320	.00134	.00056	.00024
120.0	.02518	.01039	.00436	.00184	.00078	.00033
140.0	.03152	.01335	.00570	.00244	.00104	.00044
160.0	.03812	.01660	.00721	.00313	.00135	.00058
180.0	.04486	.02008	.00890	.00391	.00170	.00073
200.0	.05162	.02374	.01073	.00479	.00211	.00091
220.0	.05830	.02753	.01269	.00575	.00256	.00112
240.0	.06479	.03139	.01476	.00679	.00306	.00135
260.0	.07104	.03527	.01691	.00789	.00360	.00160
280.0	.07697	.03912	.01912	.00906	.00417	.00187
300.0	.08255	.04289	.02134	.01026	.00479	.00217
320.0	.08773	.04653	.02357	.01150	.00543	.00248

VALUES OF CONCENTATION(C/C0) AT TIME T= 1825.0 DAYS
****

V = 1.000 A = 50.0 DL = 50.00 DT = 5.00 R = 1.00
ALFA = .0000 LAMBDA = .0000

X	Y = .0	Y = 20.0	Y = 40.0	Y = 60.0	Y = 80.0	Y = 100.0
20.0	.99014	.97812	.86961	.13257	.02288	.00660
40.0	.96864	.94316	.77018	.22641	.05057	.01532
60.0	.94778	.91013	.71285	.28181	.07964	.02580
80.0	.92394	.87709	.67573	.31608	.10748	.03757
100.0	.89844	.84574	.64954	.33878	.13278	.05010
120.0	.87227	.81663	.62961	.35463	.15510	.06292
140.0	.84616	.78980	.61347	.36609	.17455	.07565
160.0	.82064	.76509	.59977	.37454	.19139	.08802
180.0	.79605	.74229	.58771	.38084	.20596	.09986
200.0	.77258	.72119	.57681	.38552	.21857	.11106
220.0	.75034	.70159	.56676	.38894	.22951	.12158
240.0	.72934	.68333	.55736	.39137	.23902	.13140
260.0	.70957	.66628	.54848	.39299	.24729	.14053
280.0	.69097	.65031	.54001	.39394	.25450	.14899
300.0	.67349	.63531	.53192	.39434	.26077	.15682
320.0	.65705	.62120	.52414	.39429	.26624	.16404

X	Y = 120.0	Y = 140.0	Y = 160.0	Y = 180.0	Y = 200.0	Y = 220.0
20.0	.00229	.00087	.00033	.00011	.00000	-.00006
40.0	.00543	.00210	.00086	.00037	.00016	.00007
60.0	.00940	.00369	.00153	.00065	.00029	.00013
80.0	.01417	.00567	.00237	.00102	.00045	.00020
100.0	.01965	.00805	.00341	.00149	.00066	.00030
120.0	.02573	.01081	.00466	.00206	.00092	.00042
140.0	.03227	.01393	.00612	.00273	.00124	.00057
160.0	.03914	.01737	.00778	.00353	.00162	.00075
180.0	.04621	.02110	.00965	.00444	.00206	.00096
200.0	.05337	.02507	.01171	.00547	.00257	.00122
220.0	.06053	.02923	.01394	.00663	.00315	.00151
240.0	.06762	.03355	.01634	.00790	.00381	.00184
260.0	.07457	.03797	.01889	.00929	.00454	.00222
280.0	.08134	.04245	.02156	.01078	.00534	.00264
300.0	.08790	.04697	.02434	.01238	.00622	.00311
320.0	.09422	.05149	.02721	.01407	.00717	.00362

Appendix D

TDAST: A Computer Program for Evaluation of the Analytical Solution for Two–Dimensional Contaminant Transport

The program TDAST evaluates the two–dimensional analytical solute transport solution given in section 2.3.1 considering convection, dispersion, decay, and adsorption in porous media. It includes one subroutine CONC(CC0) and two functions, FUN1(X) and ERF(X), which return the integrand in (48) and error function of X, respectively. In addition, the subroutine CONC (CC0) calls for an integration routine called D01BAF which is available from the NAG Library [*Numerical Algorithms Group (NAG)*, 1981]. If the user does not have access to the NAG Library, any other accurate integration routine may be applied. The input data for this program is very short and simple. A user's guide for the program and a listing of the code are given below.

User's Guide

The following describes the necessary sequence of cards for running the program TDAST. Several problems may be solved with one run. Each problem requires one set of the following input data.

		Variable	*Description*
Card 1 (4I5)			
Col.	1–5	NUMX	Number of x positions
	6–10	NUMY	Number of y positions
	11–15	NUMT	Number of time points
	16–20	NNS	Maximum number of integrations to do to achieve convergence (see Card 7)

Card 2 (8F10.3)

One card for each eight values of x position

Col.	1–10	X(1)	Values of distance x downstream from the source, m
	11–20	X(2)	
	— —	— —	
	— —	— —	
	71–80	X(8)	

Card 3 (8F10.3)

One card for each eight values of *y* position

Col.	1–10	Y(1)	Values of *y* positions required, m
	11–20	Y(2)	
	— —	— —	
	— —	— —	
	71–80	Y(8)	

Card 4 (8F10.3)

One card for each eight values of time

Col.	1–10	T(1)	Values of time, days
	11–20	T(2)	
	— —	— —	
	— —	— —	
	71–80	T(8)	

Card 5 (4F10.4)

Col.	1–10	DL	Longitudinal dispersion coefficient, m^2/d
	11–20	DT	Transverse dispersion coefficient, m^2/d
	21–30	V	Pore water velocity, m/d
	31–40	A	Half length of source, m

Card 6 (3F10.3)

Col.	1–10	ALAM	Radioactive decay factor, d^{-1}
	11–20	R	Retardation factor
	21–30	ALFA	Decay factor of the source, d^{-1}

Card 7 (10I5)

One card for each 10 values of NS

Col.	1–5	NS(1)	First number of points for integration
	6–10	NS(2)	Second number of points for integration*
	— —	— —	...
	45–50	NS(10)	

If the value of NNS assigned in Card 1 is greater than 10 one more card is required to give the other values of NS.

Last Card

A blank card is inserted at the end of the data cards. This indicates that no further problem is to be solved at this time.

*Note: Values of NS should be chosen from the following list: 1,2,3,4,5,6,8,10,12,14,16,20,24,32,48,64.

The program calculates the integral based on the first assigned value of NS. Then it repeats the calculation of the integral based on the next assigned value of NS. If the difference between the two values is less than one percent of the magnitude of the integral calculated the second time, it stops integration and returns the value of the integral. Otherwise, it calculates the integral for the next larger value of NS. If the above criterion is not achieved and no further value of NS is available, then the program sends a message that the integral is not converging for that position, and returns the last calculated value.

```
      PROGRAM TDAST
C
C     THIS PROGRAM EVALUATES THE TWO DIMENSIONAL ANALYTICAL
C     SOLUTE TRANSPORT SOLUTION CONSIDERING CONVECTION,
C     DISPERSION, DECAY AND ADSORPTION.
C
      DIMENSION CD(100,20),X(100),Y(20),T(100)
      COMMON NS(20)
      COMMON/FAT/ NNS
      COMMON/DAT/ DL,DT,V,A
      COMMON/BAT/ALFA,ALAM,R
      COMMON/CAT/XX,YY,TT,TT0
C     ****************************************
C     READ INPUT PARAMETERS
C
C     NUMX  = NUMBER OF X POSITIONS
C     NUMY  = NUMBER OF Y POSITIONS
C     NUMT  = NUMBER OF TIME POINTS
C     NNS   = NUMBER OF INTEGRATIONS TO ACHIEVE CONVERGENCE
C     X     = X COORDINATES OF THE POINTS
C     Y     = Y COORDINATES OF THE POINTS
C     T     = TIME ELAPSED SINCE THE BEGINNING OF OPERATION
C     DL    = LONGITUDINAL DISPERSION COEF M**2/DAY
C     DT    = TRANSVERSE DISPERSION COEF M**2/DAY
C     V     = PORE WATER VELOCITY M/DAY
C     A     = HALF LENGTH OF SOURCE M
C     ALAM  = DECAY FACTOR OF THE SOLUTE 1/DAY
C     R     = RETARDATION FACTOR
C     ALFA  = DECAY FACTOR OF THE SOURCE 1/DAY
C     NS    = NUMBER OF POINTS FOR INTEGRATION
C     ****************************************
   10 READ(5,510) NUMX,NUMY,NUMT,NNS
      IF(NUMX .LT. 1)STOP
      READ(5,520) (X(I),I=1,NUMX)
      READ(5,520) (Y(I),I=1,NUMY)
      READ(5,520) (T(I),I=1,NUMT)
      READ(5,530) DL,DT,V,A
      READ(5,540) ALAM,R,ALFA
      READ(5,550) (NS(I),I=1,NNS)
C     ****************************************
C     WRITE INPUT PARAMETERS
C     ****************************************
      WRITE(6,610) V,DL,DT,A
      WRITE(6,620) ALAM,R,ALFA
      WRITE(6,630) NUMX,NUMY,NUMT
C     ****************************************
      DO 30 KK=1,NUMX
        DO 20 JJ=1,NUMY
          CD(KK,JJ)=0.0
   20     CONTINUE
```

```
30        CONTINUE
        TT0=0.0
        DO 80 I=1,NUMT
          TT=T(I)/R
          DO 50 J=1,NUMX
            XX=X(J)
            DO 40 K=1,NUMY
              YY=Y(K)
              CALL CONC(CC0)
              CD(J,K)=CC0+CD(J,K)
40            CONTINUE
50          CONTINUE
          WRITE(6,640) T(I),V,A,DL,DT,R,ALFA,ALAM
          DO 70 IX=1,NUMY,6
            LX=MIN0(IX+5,NUMY)
            WRITE(6,650)(Y(L),L=IX,LX)
            DO 60 J=1,NUMX
              WRITE(6,660) X(J),(CD(J,K),K=IX,LX)
60            CONTINUE
70          CONTINUE
          TT0=TT
80        CONTINUE
        GO TO 10
C       ****************************************
C       FORMAT STATEMENTS
C       ****************************************
510     FORMAT(4I5)
520     FORMAT(8F10.3)
530     FORMAT(4F10.4)
540     FORMAT(3F10.3 )
550     FORMAT(10I5)
C
610     FORMAT(1H1,4X,21H*CONTROL INFORMATION*,//
       1 43H  VELOCITY(M/DAY)------------------------- =F10.4/
       2 43H  LONGITUDINAL DISPERSION COEF.(M*M/DAY)- =F10.4/
       3 43H  TRANSVERSE DISPERSION COEF.(M*M/DAY)--- =F10.4/
       4 43H  HALF LENGTH OF SOURCE (M)-------------- =F10.4//)
620     FORMAT(1H ,/
       1 43H  RADIOACTIVE DECAY CONSTANT(1/DAY)------ =F10.4/
       2 43H  RETARDATION FACTOR--------------------- =F10.4/
       3 43H  SOURCE DECAY FACTOR(1/DAY)------------- =F10.4//)
630     FORMAT(1H ,42H TOTAL NUMBER OF X POSITIONS------------ =I5/
       1 43H  TOTAL NUMBER OF Y POSITIONS----------- =I5/
       2 43H  TOTAL NUMBER OF TIME POINTS----------- =I5//)
640     FORMAT(1H1,10X,39HVALUES OF CONCENTATION(C/C0) AT TIME T=
       1 ,F7.1,6H  DAYS/10X,27(2H**)//9X,2HV=,F5.3,5X,2HA=,F6.1,5X
       2 ,3HDL=,F5.2,5X,3HDT=,F5.2,5X,2HR=,F5.2// 18X,5HALFA=,F6.4,
       3 10X, 7HLAMBDA=,F6.4)
650     FORMAT(//,6X,1HX,7X,6(2HY=,F5.1,3X)/)
660     FORMAT(3X,F6.1,2X,6F10.5)
        END
```

```
      SUBROUTINE CONC(CC0)
      COMMON NS(20)
      COMMON/FAT/ NNS
      COMMON/DAT/ DL,DT,V,A
      COMMON/BAT/ALFA,ALAM,R
      COMMON/CAT/XX,YY,TT,TT0
      EXTERNAL D01BAZ,FUN1
      PI=4.*ATAN(1.)
      IFAIL=1
      I=1
      ANS1=D01BAF(D01BAZ,TT0,TT,NS(I),FUN1,IFAIL)
   90 I=I+1
      IF(I.GT.NNS) GO TO 110
      ANS2=D01BAF(D01BAZ,TT0,TT,NS(I),FUN1,IFAIL)
      ANS=ABS(ANS1-ANS2)
      ERR=ANS/AMAX1(ANS2,.1)
      IF(ERR .LT. 0.01) GO TO 100
      ANS1=ANS2
      GO TO 90
  100 CONTINUE
      CC0=ANS2
      GO TO 120
  110 CC0=ANS1
      WRITE(6,670) TT,XX,YY
  670 FORMAT(38H INTEGRAL DOES NOT CONVERGE AT TIME T=,F10.3,
     1 3X,2HX=,F10.3,2HY=,F10.3)
  120 RETURN
      END
```

```
      FUNCTION FUN1(X)
      COMMON/DAT/DL,DT,V,A
      COMMON/BAT/ALFA,ALAM,R
      COMMON/CAT/XX,YY,TT
      PI=4.*ATAN(1.)
      AA=(V*XX/(2.*DL))-(ALFA*TT)-(ALAM*R-ALFA*R+(V**2)/(4.*DL))*X-
     1 (XX**2/(4.*DL*X))
      AAA=EXP(AA)/SQRT(X**3)
      BB=(A-YY)/(2.*SQRT(DT*X))
      CC=(-A-YY)/(2.*SQRT(DT*X))
      BBB=1.-ERF(BB)
      CCC=1.-ERF(CC)
      FUN1=AAA*(CCC-BBB)*(XX/(4.*SQRT(PI*DL)))
      RETURN
      END
```

```
      FUNCTION ERF(X)
      DIMENSION D(101)
      N=100
      N1=N+1
      PI=4.*ATAN(1.)
      C=2./SQRT(PI)
      H=X/N
      DO 130 I=1,N1
        Y=(I-1)*H
130     D(I)=EXP(-Y*Y)
      E1=0.0
      DO 140 I=3,N1,2
140     E1=E1+(D(I-2)+4.*D(I-1)+D(I))*(H/3.)
      ERF=C*E1
      RETURN
      END
```

Appendix E

Tables of Dimensionless Concentration for Dispersion in Radial Flow

The following tables list dimensionless concentration C/C_o based on numerical inversion of (55) for a recharging well with dimensionless well radius $r_{Dw} = 0.1$. C/C_o is given as a function of dimensionless time for different values of dimensionless radius.

DIMENSIONLESS CONCENTRATION C/C0 FOR

RDW= .10

TD\RD	10.	20.	30.	40.	50.	60.	70.	80.	90.
200.	1.00	.48	.00	.00	.00	.00	.00	.00	.00
400.	1.00	.97	.33	.00	.00	.00	.00	.00	.00
600.	1.00	1.00	.81	.13	.00	.00	.00	.00	.00
800.	1.00	1.00	.97	.49	.01	.00	.00	.00	.00
1000.	1.00	1.00	1.00	.77	.18	.00	.00	.00	.00
1200.	1.00	1.00	1.00	.93	.43	.01	.00	.00	.00
1400.	1.00	1.00	1.00	.99	.66	.13	.00	.00	.00
1600.	1.00	1.00	1.00	1.00	.82	.31	.00	.00	.00
1800.	1.00	1.00	1.00	1.00	.92	.49	.06	.00	.00
2000.	1.00	1.00	1.00	1.00	.97	.66	.18	.00	.00
2200.	1.00	1.00	1.00	1.00	1.00	.79	.32	.01	.00
2400.	1.00	1.00	1.00	1.00	1.00	.88	.46	.07	.00
2600.	1.00	1.00	1.00	1.00	1.00	.94	.59	.16	.00
2800.	1.00	1.00	1.00	1.00	1.00	.98	.71	.27	.00
3200.	1.00	1.00	1.00	1.00	1.00	1.00	.87	.50	.12
3400.	1.00	1.00	1.00	1.00	1.00	1.00	.92	.60	.20

TD\RD	70.	80.	90.	100.	110.	120.	130.	140.	150.
3400.	.92	.60	.20	.00	.00	.01	.01	.00	.00
3800.	.98	.77	.39	.07	.00	.00	.00	.00	.00
4200.	1.00	.89	.56	.20	.00	.00	.00	.00	.00
4600.	1.00	.96	.71	.35	.06	.00	.00	.00	.00
5000.	1.00	.99	.83	.50	.17	.00	.00	.00	.00
5400.	1.00	1.00	.91	.64	.29	.04	.00	.00	.00
5800.	1.00	1.00	.96	.75	.43	.13	.00	.00	.00
6200.	1.00	1.00	.99	.84	.55	.23	.02	.00	.00
6600.	1.00	1.00	1.00	.91	.66	.34	.08	.00	.00
7000.	1.00	1.00	1.00	.95	.75	.45	.16	.00	.00
7400.	1.00	1.00	1.00	.98	.83	.56	.26	.04	.00
7800.	1.00	1.00	1.00	1.00	.89	.65	.35	.10	.00
8200.	1.00	1.00	1.00	1.00	.94	.73	.45	.18	.00
8600.	1.00	1.00	1.00	1.00	.97	.80	.54	.26	.05
9000.	1.00	1.00	1.00	1.00	.99	.86	.62	.34	.11
9400.	1.00	1.00	1.00	1.00	1.00	.91	.70	.43	.17

Appendix F

LTIRD: A Computer Program for a Semianalytical Solution to Radial Dispersion in Porous Media

The program LTIRD calculates the dimensionless concentration of a particular solute, injected into an aquifer, as a function of dimensionless time for different values of dimensionless radius. The evaluation is based on the numerical Laplace transform inversion of (55). The program was originally written by *Moench and Ogata* [1981] and has been modified for this work. It includes a routine called LINV which is used for the numerical inversion. The input data for this program is very short and simple. A user's guide for the program and a listing of the code are given below.

User's Guide

The following describes the necessary sequence of cards for running the program LTIRD.

		Variable	*Description*
Card 1 (2I5,D10.4)			
Col.	1–5	NUMR	Number of dimensionless radii
	6–10	NUMT	Number of dimensionless times
	11–20	RDW	Dimensionless radius of the well

Card 2 (6D10.4)

One card for each six values of dimensionless radius

Col.	1–10	RD(1)	Values of dimensionless radius
	11–20	RD(2)	
	— —	— —	
	— —	— —	
	51–60	RD(6)	

Card 3 (6D10.4)

One card for each six values of dimensionless time

Col.	1–10	TD(1)	Values of dimensionless time
	11–20	TD(2)	
	— —	— —	
	— —	— —	
	51–60	TD(6)	

```
      PROGRAM LTIRD
C     LAPLACE TRANSFORM INVERSION FOR RADIAL DISPERSION
C     ************************************************
C
C     NUMR  = NUMBER OF DIMENSIONLESS RADII
C     NUMT  = NUMBER OF DIMENSIONLESS TIMES
C     RDW   = DIMENSIONLESS RADIUS OF THE WELL
C     RD    = DIMENSIONLESS RADII
C     TD    = DIMENSIONLESS TIMES
C     N     = AN EVEN NUMBER GOVERNING THE ACCURACY OF
C              THE INVERSION
C     NN    = NUMBER OF TERMS IN THE AIRY SERIES EXPANSION
C     G     = ARRAY OF FACTORIALS
      COMMON N,G(50),V(50),H(25)
      DIMENSION C(20),RD(60),TD(60),FA(9)
      DOUBLE PRECISION G,V,H,RD,TD,FA,A,AA,B,BB,BOA,
     1 CC,FF,RDW,TAU,RHO,DD,XP,S,SS,ARG,C,PART1,
     2 PART2,AS,BS,SUMN,SUMD,CBAR,E2,AK,BK,E3
      N=14
      CALL LINV
      NN=5
      AA=1.D0/3
      BB=2*AA
      CC=DLOG(2.D0)
      FF=.25
10    READ(5,510) NUMR,NUMT,RDW
      IF (NUMR.EQ.0) STOP
      READ(5,520) (RD(I), I=1,NUMR)
      READ(5,520) (TD(I), I=1,NUMT)
      WRITE(6,610) RDW
      DO 30 I=2,NN
        K1=2*I-1
        E2=1
        JN=2*I-2
        DO 20 J=1,JN
          E2=K1*E2
          K1=K1+2
20        CONTINUE
        AK=E2
        E3=216.D0**(I-1)
        BK=E3*G(I-1)
        C(I)=AK/BK
30      CONTINUE
       DO 80 IR1=1,NUMR,9
         IR9=MIN0(IR1+8,NUMR)
         WRITE(6,620) (RD(IR),IR=IR1,IR9)
         WRITE(6,630)
         DO 70 IT=1,NUMT
           TAU=TD(IT)
           SS=CC/TAU
           IRK=0
           DO 60 IR=IR1,IR9
             IRK=IRK+1
             RHO=RD(IR)
             DD=(RHO-RDW)/2
             XP=0
             DO 50 L=1,N
               S=L*SS
               A=RHO*S+FF
```

```
                    B=RDW*S+FF
                    BOA=B/A
                    ARG=DD-(BB/S)*(A**(1/BB)-B**(1/BB))
                    PART1=(BOA)**FF*DEXP(ARG)/S
                    AS=BB*A**(1/BB)/S
                    BS=BB*B**(1/BB)/S
                    SUMN=0
                    SUMD=0
                    NNN=NN-1
                    DO 40 K=1,NNN
                       SUMN=(-1)**K*C(K+1)/AS**K+SUMN
                       SUMD=(-1)**K*C(K+1)/BS**K+SUMD
 40                    CONTINUE
                    PART2=(1+SUMN)/(1+SUMD)
                    CBAR=PART1+PART2
                    XP=XP+V(L)*CBAR
 50                 CONTINUE
                 FA(IRK)=SS*XP
                 IF (FA(IRK).GT.1) FA(IRK)=1
                 IF (FA(IRK).LT.1.E-5) FA(IRK)=1.E-5
 60              CONTINUE
              WRITE(6,640) TAU,(FA(IR),IR=1,IRK)
 70           CONTINUE
           WRITE(6,650)
 80        CONTINUE
         GO TO 10
510      FORMAT(2I5,E10.4)
520      FORMAT(6E10.4)
610      FORMAT(1H1,20X,36HDIMENSIONLESS CONCENTRATION C/C0 FOR
        1     /1H0,34X,4HRDW=,F4.2/)
620      FORMAT(9H0   TD\RD,9(F6.0,1X))
630      FORMAT(1X,7(10(1H-)))
640      FORMAT(1X,F7.0,9F7.2)
650      FORMAT(1H0/)
         END
```

```
      SUBROUTINE LINV
      COMMON N,G(50),V(50),H(25)
      DOUBLE PRECISION G,V,H,FI,SN
      G(1)=1
      NH=N/2
      DO 90 I=2,N
        G(I)=G(I-1)*I
 90     CONTINUE
      H(1)=2/G(NH-1)
      DO 100 I=3,NH
        FI=I-1
        H(I-1)=FI**NH*G(2*I-2)/(G(NH-I+1)*G(I-1)*G(I-2))
100     CONTINUE
      FI=NH
      H(NH)=FI**NH*G(2*NH)/(G(NH)*G(NH-1))
      SN=2*MOD(NH,2)-1
      DO 140 I=1,N
        V(I)=0
        K1=(I+1)/2
        K2=MIN0(I,NH)
        DO 130 K=K1,K2
          IF (2*K.EQ.I) GO TO 110
          IF (I.EQ.K) GO TO 120
          V(I)=V(I)+H(K)/(G(I-K)*G(2*K-I))
          GO TO 130
110       V(I)=V(I)+H(K)/G(I-K)
          GO TO 130
120       V(I)=V(I)+H(K)/G(2*K-I)
130       CONTINUE
        V(I)=SN*V(I)
        SN=-SN
        WRITE(6,660) I,V(I)
140     CONTINUE
      RETURN
660   FORMAT(3H0V(,I2,2H)=,D26.20)
      END
```

Appendix G: Tables of Error Function

X	ERF	X	ERF	X	ERF	X	ERF
.01	.011283415556	.46	.484655390002	.91	.801882825766	1.36	.945561436561
.02	.022564574692	.47	.493745050886	.92	.806767721548	1.37	.947312398035
.03	.033841222342	.48	.502749670695	.93	.811563558585	1.38	.949016035256
.04	.045111106145	.49	.511668261189	.94	.816271018976	1.39	.950673295805
.05	.056371977797	.50	.520499877813	.95	.820890807273	1.40	.952285119763
.06	.067621594393	.51	.529243619841	.96	.825423649644	1.41	.953852439360
.07	.078857719771	.52	.537898630479	.97	.829870293036	1.42	.955376178641
.08	.090078125841	.53	.546464096935	.98	.834231504340	1.43	.956857253145
.09	.101280593915	.54	.554939250456	.99	.838508069555	1.44	.958296569601
.10	.112462916018	.55	.563323366325	1.00	.842700792950	1.45	.959695025637
.11	.123622896199	.56	.571615763824	1.01	.846810496228	1.46	.961053509513
.12	.134758351820	.57	.579815806164	1.02	.850838017701	1.47	.962372899854
.13	.145867114836	.58	.587922900382	1.03	.854784211454	1.48	.963654065414
.14	.156947033063	.59	.595936497198	1.04	.858649946527	1.49	.964897864843
.15	.167995971427	.60	.603856090848	1.05	.862436106090	1.50	.966105146475
.16	.17901181319	.61	.611681218876	1.06	.866143586635	1.51	.967276748129
.17	.189992461202	.62	.619411461899	1.07	.869773297164	1.52	.968413496920
.18	.200935839019	.63	.627046443338	1.08	.873326158388	1.53	.969516209093
.19	.211839892158	.64	.634585829122	1.09	.876803101938	1.54	.970585689861
.20	.222702589210	.65	.642029327356	1.10	.880205069574	1.55	.971622733262
.21	.233521922982	.66	.649376687963	1.11	.883533012415	1.56	.972628122027
.22	.244295911599	.67	.656627702300	1.12	.886787890165	1.57	.973602627462
.23	.255022599592	.68	.663782202741	1.13	.889970670363	1.58	.974547009343
.24	.265700058954	.69	.670840062235	1.14	.893082327630	1.59	.975462015822
.25	.276326390168	.70	.677801193837	1.15	.896123842937	1.60	.976348383345
.26	.286899723216	.71	.684665550217	1.16	.899096202880	1.61	.977206836583
.27	.297418218547	.72	.691433123139	1.17	.902000398966	1.62	.978038088373
.28	.307880068029	.73	.698103942917	1.18	.904837426915	1.63	.978842839674
.29	.318283495861	.74	.704678077855	1.19	.907608285972	1.64	.979621779524
.30	.328626759459	.75	.711155633654	1.20	.910313978230	1.65	.980375585023
.31	.338908150311	.76	.717536752806	1.21	.912955507973	1.66	.981104921311
.32	.349125994796	.77	.723821613965	1.22	.915533881027	1.67	.981810441565
.33	.359278654974	.78	.730010431299	1.23	.918050104127	1.68	.982492787002
.34	.369364529345	.79	.736103453821	1.24	.920505184299	1.69	.983152586895
.35	.379382053562	.80	.742100964708	1.25	.922900128256	1.70	.983790458591
.36	.389329701129	.81	.748003280598	1.26	.925235941810	1.71	.984407007545
.37	.399205984043	.82	.753810750875	1.27	.927513629295	1.72	.985002827359
.38	.409009453420	.83	.759523756938	1.28	.929734193014	1.73	.985578499828
.39	.418738700070	.84	.765142711455	1.29	.931898632689	1.74	.986134594997
.40	.428392355047	.85	.770668057608	1.30	.934007944941	1.75	.986671671219
.41	.437969090155	.86	.776100268324	1.31	.936063122773	1.76	.987190275231
.42	.447467618426	.87	.781439845491	1.32	.938065155079	1.77	.987690942224
.43	.456886694550	.88	.786687319174	1.33	.940015026158	1.78	.988174195930
.44	.466225115278	.89	.791843246814	1.34	.941913715258	1.79	.988640548706
.45	.475481719787	.90	.796908212423	1.35	.943762196123	1.80	.989090501636

X	ERF	X	ERF	X	ERF	X	ERF
1.81	.989524544622	2.26	.998607121117	2.71	.999873162073	3.16	.999992138259
1.82	.989943156497	2.27	.998673872484	2.72	.999880261487	3.17	.999992641873
1.83	.990346805133	2.28	.998737661150	2.73	.999886985015	3.18	.999993114550
1.84	.990735947553	2.29	.998798606423	2.74	.999893351286	3.19	.999993558102
1.85	.991111030056	2.30	.998856823403	2.75	.999899378078	3.20	.999993974239
1.86	.991472488336	2.31	.998912423104	2.76	.999905082352	3.21	.999994364578
1.87	.991820747611	2.32	.998965512573	2.77	.999910480288	3.22	.999994730645
1.88	.992156222755	2.33	.999016195007	2.78	.999915587316	3.23	.999995073881
1.89	.992479318433	2.34	.999064569861	2.79	.999920418147	3.24	.999995395645
1.90	.992790429235	2.35	.999110732968	2.80	.999924986805	3.25	.999995697221
1.91	.993089939820	2.36	.999154776641	2.81	.999929306654	3.26	.999995979817
1.92	.993378225057	2.37	.999196789784	2.82	.999933390427	3.27	.999996244577
1.93	.993655650170	2.38	.999236857995	2.83	.999937250254	3.28	.999996492576
1.94	.993922570889	2.39	.999275063670	2.84	.999940897685	3.29	.999996724828
1.95	.994179333592	2.40	.999311486103	2.85	.999944343720	3.30	.999996942290
1.96	.994426275465	2.41	.999346201583	2.86	.999947598827	3.31	.999997145864
1.97	.994663724647	2.42	.999379283488	2.87	.999950672970	3.32	.999997336397
1.98	.994892000387	2.43	.999410802386	2.88	.999953575628	3.33	.999997514690
1.99	.995111413200	2.44	.999440826116	2.89	.999956315820	3.34	.999997681496
2.00	.995322265019	2.45	.999469419888	2.90	.999958902122	3.35	.999997837523
2.01	.995524849355	2.46	.999496646359	2.91	.999961342688	3.36	.999997983439
2.02	.995719451452	2.47	.999522565729	2.92	.999963645269	3.37	.999998119872
2.03	.995906348444	2.48	.999547235814	2.93	.999965817233	3.38	.999998247413
2.04	.996085809512	2.49	.999570712132	2.94	.999967865580	3.39	.999998366617
2.05	.996258096044	2.50	.999593047983	2.95	.999969796958	3.40	.999998478007
2.06	.996423461790	2.51	.999614294518	2.96	.999971617683	3.41	.999998582074
2.07	.996582153017	2.52	.999634500823	2.97	.999973333751	3.42	.999998679281
2.08	.996734408670	2.53	.999653713985	2.98	.999974950855	3.43	.999998770061
2.09	.996880460524	2.54	.999671979162	2.99	.999976474397	3.44	.999998854822
2.10	.997020533344	2.55	.999689339657	3.00	.999977909503	3.45	.999998933948
2.11	.997154845031	2.56	.999705836980	3.01	.999979261036	3.46	.999999007799
2.12	.997283606785	2.57	.999721510911	3.02	.999980533609	3.47	.999999076712
2.13	.997407023251	2.58	.999736399570	3.03	.999981731595	3.48	.999999141004
2.14	.997525292671	2.59	.999750539471	3.04	.999982859140	3.49	.999999200975
2.15	.997638607037	2.60	.999763965583	3.05	.999983920174	3.50	.999999256902
2.16	.997747152237	2.61	.999776711391	3.06	.999984918421	3.51	.999999309048
2.17	.997851108202	2.62	.999788808944	3.07	.999985857407	3.52	.999999357659
2.18	.997950649053	2.63	.999800288918	3.08	.999986740475	3.53	.999999402965
2.19	.998045943243	2.64	.999811180661	3.09	.999987570788	3.54	.999999445184
2.20	.998137153702	2.65	.999821512248	3.10	.999988351343	3.55	.999999484516
2.21	.998224437976	2.66	.999831310527	3.11	.999989084973	3.56	.999999521153
2.22	.998307948365	2.67	.999840601169	3.12	.999989774362	3.57	.999999555272
2.23	.998387832062	2.68	.999849408712	3.13	.999990422049	3.58	.999999587040
2.24	.998464231285	2.69	.999857756607	3.14	.999991030434	3.59	.999999616613
2.25	.998537283413	2.70	.999865667260	3.15	.999991601789	3.60	.999999644137

X	ERF	X	ERF	X	ERF	X	ERF
3.61	.999999669749	4.06	.999999990627	4.51	.999999999821	4.96	.999999999998
3.62	.999999693577	4.07	.999999991379	4.52	.999999999837	4.97	.999999999998
3.63	.999999715741	4.08	.999999992072	4.53	.999999999851	4.98	.999999999998
3.64	.999999736353	4.09	.999999992711	4.54	.999999999864	4.99	.999999999998
3.65	.999999755517	4.10	.999999993300	4.55	.999999999876	5.00	.999999999998
3.66	.999999773333	4.11	.999999993842	4.56	.999999999887	5.01	.999999999999
3.67	.999999789891	4.12	.999999994342	4.57	.999999999897	5.02	.999999999999
3.68	.999999805277	4.13	.999999994802	4.58	.999999999906	5.03	.999999999999
3.69	.999999819571	4.14	.999999995225	4.59	.999999999915	5.04	.999999999999
3.70	.999999832849	4.15	.999999995615	4.60	.999999999999	3.05	.999999999999
3.71	.999999845179	4.16	.999999995974	4.61	.999999999929	5.06	.999999999999
3.72	.999999856628	4.17	.999999996304	4.62	.999999999936	5.07	.999999999999
3.73	.999999867256	4.18	.999999996608	4.63	.999999999942	5.08	.999999999999
3.74	.999999877120	4.19	.999999996888	4.64	.999999999947	5.09	.999999999999
3.75	.999999886273	4.20	.999999997145	4.65	.999999999952	5.10	.999999999999
3.76	.999999894764	4.21	.999999997381	4.66	.999999999956	5.11	1.00000000000
3.77	.999999902641	4.22	.999999997598	4.67	.999999999960	5.12	1.00000000000
3.78	.999999909945	4.23	.999999997798	4.68	.999999999964	5.13	1.00000000000
3.79	.999999916718	4.24	.999999997981	4.69	.999999999967	5.14	1.00000000000
3.80	.999999922996	4.25	.999999998149	4.70	.999999999970	5.15	1.00000000000
3.81	.999999928815	4.26	.999999998304	4.71	.999999999973	5.16	1.00000000000
3.82	.999999934207	4.27	.999999998446	4.72	.999999999975	5.17	1.00000000000
3.83	.999999939202	4.28	.999999998577	4.73	.999999999978	5.18	1.00000000000
3.84	.999999943829	4.29	.999999998697	4.74	.999999999980	5.19	1.00000000000
3.85	.999999948114	4.30	.999999998807	4.75	.999999999982	5.20	1.00000000000
3.86	.999999952081	4.31	.999999998907	4.76	.999999999983	5.21	1.00000000000
3.87	.999999955754	4.32	.999999999000	4.77	.999999999985	5.22	1.00000000000
3.88	.999999959153	4.33	.999999999085	4.78	.999999999986	5.23	1.00000000000
3.89	.999999962298	4.34	.999999999163	4.79	.999999999987	5.24	1.00000000000
3.90	.999999965208	4.35	.999999999234	4.80	.999999999989	5.25	1.00000000000
3.91	.999999967899	4.36	.999999999299	4.81	.999999999990	5.26	1.00000000000
3.92	.999999970388	4.37	.999999999359	4.82	.999999999991	5.27	1.00000000000
3.93	.999999972690	4.38	.999999999414	4.83	.999999999992	5.28	1.00000000000
3.94	.999999974817	4.39	.999999999465	4.84	.999999999992	5.29	1.00000000000
3.95	.999999976783	4.40	.999999999511	4.85	.999999999993	5.30	1.00000000000
3.96	.999999978600	4.41	.999999999553	4.86	.999999999994	5.31	1.00000000000
3.97	.999999980279	4.42	.999999999592	4.87	.999999999994	5.32	1.00000000000
3.98	.999999981829	4.43	.999999999627	4.88	.999999999995	5.33	1.00000000000
3.99	.999999983261	4.44	.999999999659	4.89	.999999999995	5.34	1.00000000000
4.00	.999999984583	4.45	.999999999689	4.90	.999999999996	5.35	1.00000000000
4.01	.999999985803	4.46	.999999999716	4.91	.999999999996	5.36	1.00000000000
4.02	.999999986929	4.47	.999999999741	4.92	.999999999997	5.37	1.00000000000
4.03	.999999987969	4.48	.999999999764	4.93	.999999999997	5.38	1.00000000000
4.04	.999999988927	4.49	.999999999784	4.94	.999999999997	5.39	1.00000000000
4.05	.999999989812	4.50	.999999999803	4.95	.999999999997	5.40	1.00000000000

Appendix H

RESSQ: A Computer Program for Semianalytical Contaminant Transport

The computer program RESSQ calculates two-dimensional contaminant transport by advection and adsorption (no dispersion or diffusion) in a homogeneous, isotropic confined aquifer of uniform thickness when regional flow, sources, and sinks create a steady state flow field. Recharge wells and ponds act as sources and pumping wells act as sinks. RESSQ calculates the streamline pattern in the aquifer (subroutine FLOW), the location of contaminant fronts around sources at various times (subroutine FLOW), and the variation of contaminant concentration with time at sinks (subroutine CONCEN). RESSQ was developed at the Lawrence Berkeley Laboratory based on a solution procedure used by *Gringarten and Sauty* [1975*a*,*b*]. A user's guide for the program and a listing of the code are given below.

User's Guide

The following describes the input required for RESSQ.

		Variable	*Description*
Card 1	(20A4)		
Col.	1–80	TITLE	Any 80 characters
Card 2	(2I5, 2E10.4, A10, I5)		
Col.	1–5	NWI	Number of injection wells (>0, see note 7 below)
	6–10	NWP	Number of production wells
	11–20	C0	Ambient concentration in aquifer
	21–30	CD	Default injection concentration
	31–40	UNITC	Units of concentration (default "Percent")
	41–45	NSYST	System of units (1 indicates CGS, blank indicates practical). The expected unit is given in parentheses after each variable name as (CGS unit or practical unit).
Card 3	(5E10.4)		
Col.	1–10	HEIGHT	Thickness of aquifer (cm or m)
	11–20	POR	Porosity (range 0–1)
	21–30	V0	Pore water velocity of uniform regional flow (cm/s or m/yr)
	31–40	ALPHA	Direction of regional flow (in degrees, measured counterclockwise from positive x axis)
	41–50	ADSORB	Adsorption capacity of rock matrix, equals $(1-1/R)$, where R is retardation factor from (7). (range 0–1, 0 indicates no adsorption, 1 indicates total adsorption)

Card 4 (5A2, 6E10.4, 3I2)

This card is repeated for each well — injection wells first.
C, BETA1, NSL, and ITR are ignored for production wells.

Col.			
Col.	1–10	NAMEW	Name of the well (maximum 10 characters)
	11–20 21–30	XW YW	Coordinates of the well (cm or m)
	31–40	QW	Flow rate into/from the well, non-negative (cm^3/s or m^3/h). For injection wells, QW=0 indicates a point from which to trace a single streamline (see note 7 below).
	41–50	RADW	Radius of the well (cm or m) (default 7.5 cm)
	51–60	C	Injection concentration in units given by UNITC (default CD)
	61–70	BETA1	Angle (in degrees) at which the first streamline calculated (also the first one plotted) leaves the well. The remaining streamlines depart at equally spaced angles measured counter-clockwise from the first (0 indicates positive x axis).
	71–72	NSL	Number of streamlines calculated for the well (–1 indicates no lines), forced to be 1 if QW=0
	73–74	ITR	Ratio of NSL to the number of streamlines plotted (–1 indicates no lines plotted from the well) Defaults: NSL=40, ITR=4, 10 lines plotted
	75–76	INDW	Flag: INDW=–1 suppresses: plot of fronts in the case of an injection well, study of concentration in the case of a production well.

Card 5 (I5, 5X, 7E10.4)

Col.			
Col.	1–5	NFRNTS	Number of fronts to be calculated around each injection well (maximum 7)
	11–20	DATE(1)	Times at which fronts are calculated (seconds or years)
	21–30	DATE(2)	
	— —	— —	
	— —	— —	
	71–80	DATE(7)	

Card 6 (2E10.4, 2I5)

1–10	TMAX	Period of study — maximum amount of time for calculating the trace of a streamline (seconds or years)
11–20	DL	Step length — spatial increment used to trace out streamlines (cm or m) (default (XMAX–XMIN)/200).
21–25	NTL	Flag: NTL = –1 suppresses plot of streamlines
26–30	NTF	Flag: NTF = –1 suppresses plot of fronts

Card 7 (4E10.4)

1–10	XMIN	Limits of the area studied (cm or m)
11–20	XMAX	
21–30	YMIN	
31–4)	YMAX	

To do a series of runs, cards 2 through 7 may be repeated as often as necessary. The program ends when an end–of–file mark is found instead of card 2.

The program RESSQ produces two output files. The first file, called OUTPUT, lists the problem input, the final location and arrival time of the streamlines leaving each injection well, and the variation of concentration at each production well. The second file, called TAPE7, contains data suitable for plotting: the plot area limits, well locations, streamline coordinates, front locations, and production well concentration versus time data. An example of OUTPUT is given in Tables 8 and 9. An example of TAPE7 is given in Tables 10 and 11. A few notes on the use of RESSQ are given below, followed by a listing of the code.

1. If the total flow rate from all injection wells does not equal the total flow rate from all production wells, then, strictly speaking, a steady state flow field, as required by RESSQ, cannot be achieved. However, for large values of time one may assume that quasi–steady flow prevails, thus allowing RESSQ to be used.
2. The step length used to trace out streamlines (DL) affects the accuracy of the solution and the cost of running the program. A smaller step yields a smoother streamline but increases the number of calculations that must be done. For a particular problem, several step lengths may be tried for short periods of study to determine the step length that produces the desired accuracy with minimum cost.
3. Because there is a fixed amount of memory allocated for each streamline, storage of long streamlines may require the elimination of alternate points during the calculation with subroutine WEED. If WEED is called repeatedly, points near the beginning of the streamline may be very far apart. Shortening the period studied will eliminate some of the calls to WEED and allow more early points to be saved. Alternatively, increasing the dimension of the array A (NDIMA) in the program (lines 11 and 46) will increase the amount of memory available for each streamline.
4. In addition to modeling recharge or injection wells as point sources, RESSQ can model constant head ponds as finite radius sources. The input required is the same as for injection wells (card 4) with the radius of the pond used for RADW.
5. RESSQ can also include a linear no–flow or constant potential boundary using the method of images. A boundary is represented by adding an image well for each real well in the problem, with the boundary located on the perpendicular bisector of the line connecting each real well/image well pair. For a no–flow boundary the real and image wells have the same flow rate, that is, either both are injection or both are production wells. Since there is no flow through an impervious boundary, the only regional flow allowed in this case is parallel to the boundary. For a constant potential boundary the real well/image well pairs have flow rates equal in magnitude and opposite in sign, that is, one is an injection well and the other is a production well. In this case, since the boundary must be an equipotential, the only regional flow allowed is perpendicular to the boundary.
6. RESSQ calculates the locations of contaminant fronts around injection wells for a desired series of times, as specified in card 5. For a one–injection well problem the time t_1 front outlines the region of the aquifer into which contaminant flows during time t_1. By symmetry this front also outlines the region from which contaminant would be extracted in a time t_1 if the injection well were replaced by a production well with the same flow rate and regional flow reversed. The same principle applies for multiwell problems. Thus to use RESSQ to outline the regions from which contam-

inant is produced in a time t_1, input all sinks as sources and vice versa, add 180° to the direction of regional flow, and request that front location be calculated at time t_1.

7. As noted in the input description for card 2, the number of injection wells, NWI, must be greater than zero. This is because injection wells act as streamline starting points, so without injection wells no streamline pattern can be drawn. To allow greater flexibility in presenting streamline patterns, the NWI injection wells may consist of some or all wells with a flow rate of zero. Zero–flow rate wells do not affect the velocity field, of course, but provide starting points for streamlines whose paths may help explicate the velocity field created by regional flow and nonzero–flow rate sources and sinks present.

8. In general, either of the following two approaches may be taken when using RESSQ to calculate streamline patterns.

The "qualitative" approach involves using the streamlines drawn by RESSQ to illustrate the direction of contaminant flow without providing information on the magnitude of the flow. Any convenient number of streamlines may leave each nonzero–flow rate source, and zero–flow rate wells may be placed at arbitrary points of interest. The spacing of the resulting streamlines is not indicative of the magnitude of flow. The qualitative approach provides a convenient way to focus on the direction of flow from a particular point. For instance, if a contaminant plume has been located in an aquifer, then the streamlines from zero–flow rate wells placed around the outline of the plume can show where the plume moves due to regional flow, use of local water supply wells, or aquifer restoration operations.

The quantitative approach considers streamlines in which the density of streamlines is proportional to the magnitude of flow everywhere. Thus the number of streamlines plotted leaving each injection well, N = NSL/ITR, should be proportional to the flow rate of that well:

$$\frac{Q_1}{N_1} \approx \frac{Q_2}{N_2} \tag{H1}$$

where N_2 = is the number of streamlines plotted leaving an injection well with flow rate Q_2 and N_1 is the number of streamlines plotted leaving an injection well with flow rate Q_1. Because N_1 and N_2 must be integers, (H1) is only exact when $N_1(Q_2/Q_1)$ is an integer. Larger values of N improve the accuracy of (H1). For greatest accuracy, ITR should be equal to 1; thus N = NSL for each well. Streamlines showing regional flow can be drawn by placing a row of zero–flow rate wells perpendicular to the direction of regional flow at a distance far from sources and sinks. Since each zero–flow rate well generates one streamline, the spacing between the wells, D, measured perpendicular to the regional flow must be given by

$$D = \frac{Q_1}{N_1} \frac{1}{bU} \tag{H2}$$

where Q_1/N_1 is the ratio of flow rate to the number of streamlines plotted for each source, b is aquifer thickness, and U is the Darcy velocity of regional flow. If there are no sources, just regional flow and sinks, then Q_1/N_1 may be chosen as the desired ratio of flow rate to number of streamlines captured for each production well. A simple auxiliary program called ZQWELL may be used to calculate the zero–flow rate well locations and print them in a format suitable for input to RESSQ. The input format for ZQWELL is given below. Figure H1 provides a sketch illustrating the parameters.

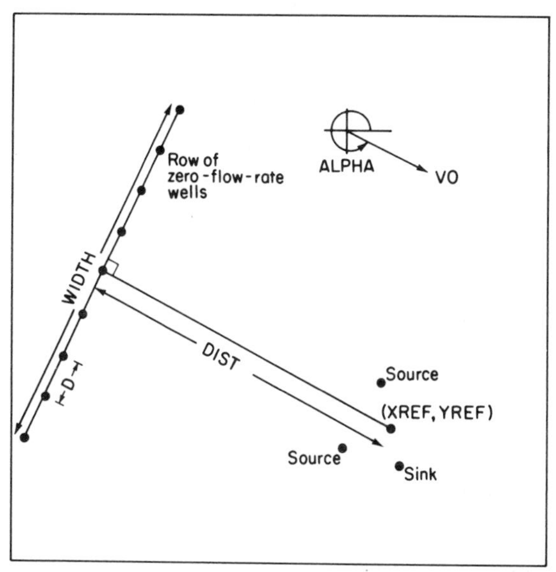

Fig. H1. Sketch showing parameters used for program ZQWELL.

		Variable	*Description*
Card 1	(4E10.4)		
Col.	1–10	XREF	(x,y) coordinates of an arbitrary reference
	11–20	YREF	point near the sources and sinks (cm or m)
	21–30	DIST	Distance from (XREF, YREF) to row of zero–flow rate wells (cm or m)
	31–40	WIDTH	Width of row of zero–flow rate wells (cm or m)
Card 2	(5E10.4, 2I2)		
Col.	1–10	H	Aquifer thickness (cm or m)
	11–20	PHI	Porosity (range 0–1)
	21–30	V0	Pore water velocity of regional flow (cm/s or m/yr)
	31–40	ALPHA	Direction of regional flow (degrees) (see Figure H1)
	41–50	Q1	Flow rate of first source (cm^3/s or m^3/h)
	51–52	N1	Number of streamlines plotted for first source (NSL/ITR)
	53–54	NSYST	System of units (1 indicates CGS, blank indicates practical)

DIST must be large enough so that near the zero–flow rate wells the streamlines are essentially parallel. Table H1 shows the input and output for ZQWELL that were used to produce the regional flow streamlines of Figure 20. A listing of ZQWELL follows the listing of RESSQ.

With N for each injection well and D calculated according to (H1) and (H2) the density of streamlines everywhere will be proportional to the magnitude of flow. Thus

TABLE H1. Input and output for program ZQWELL for the regional flow streamlines of Figure 20

INPUT

Card	Column	Entry	Description
1	1–10	0.	x coordinate of reference point for row of zero–flow rate wells (m)
	11–20	0.	y coordinate of reference point (m)
	21–30	3000.	Distance between reference point and row of wells (m)
	31–40	3000.	Width of row of wells (m)
2	1–10	10.	Aquifer thickness (m)
	11–20	.25	Porosity
	21–30	50.	Pore water velocity of regional flow (m/yr)
	31–40	45.	Angle of regional flow with respect to positive x axis (degrees)
	41–50	50.	Flow rate of injection well (m^3/h)
	51–52	15	Number of streamlines plotted (NSL/ITR=45/3)
	53–54		Blank for practical units

OUTPUT

Well Name	XW (m)	YW (m)	QW (m^3/hr)
ZQWELL 1	−3196.	−1047.	0.
ZQWELL 2	−3030.	−1212.	0.
ZQWELL 3	−2865.	−1378	0.
ZQWELL 4	−2700.	−1543.	0.
ZQWELL 5	−2535.	−1708.	0.
ZQWELL 6	−2369.	−1873.	0.
ZQWELL 7	−2204.	−2039.	0.
ZQWELL 8	−2039.	−2204.	0.
ZQWELL 9	−1873.	−2369.	0.
ZQWELL 10	−1708.	−2535.	0.
ZQWELL 11	−1543.	−2700.	0.
ZQWELL 12	−1378.	−2865.	0.
ZQWELL 13	−1212.	−3030.	0.
ZQWELL 14	−1047.	−3196.	0.

the number of streamlines eventually captured by a production well with flow rate Q_p should be given by

$$N_p \approx \frac{N_1}{Q_1} Q_p \tag{H3}$$

which is exact when $N_1(Q_p/Q_1)$ is an integer. However, due to various approximations used by RESSQ (e.g., (H1) and (H3) are approximations; practically, DIST cannot be large enough so that streamlines are exactly parallel; only a finite number of streamlines are calculated so there is an element of arbitrariness in their placement), N_p may not be correct. That is, a production well may just miss a streamline it should capture. Such a streamline pattern can be fine tuned by changing N for each injection well, increasing DIST, or by resetting the arbitrary parameters for each injection well: BETA1, the angle the first streamline leaves a source, and (XREF, YREF), the reference coordinates for the row of zero–flow rate wells. A typical BETA1 adjustment is 1/3 to 1/2 of 360/NSL. A typical (XREF, YREF) adjustment is 1/2 of D, perpendicular to the direction of regional flow.

For a quasi-steady state problem in which the total flow rate from all production wells is greater than the total flow rate from all injection wells and regional velocity is zero, too few streamlines are available for the quantitative approach to yield correct values of N_p for all production wells. To make a quantitative streamline picture, the technique described in note 6 may be used: all sinks are input as sources and vice versa. Fronts around injection wells may be added by running the original problem (i.e., sources input as sources, sinks as sinks) with NTL=–1 (no streamlines plotted), then superimposing the two plots.

```
      PROGRAM RESSQ
C              (INPUT,OUTPUT,TAPE7,TAPE5=INPUT,TAPE6=OUTPUT)
      COMMON/WELLS/NWI,NWP,NWELL,IP1,CD
      COMMON/IOUNIT/UNITL,UNITQ,UNITV,UNITT,UNITR,UNISYS,UNITC,
     1              COEFL,COEFQ,COEFV,COEFT,COEFR,NSYST
      CHARACTER*10 UNITL,UNITQ,UNITV,UNITT,UNITR,UNISYS,UNITC
      CHARACTER*10 UNIT(6),UNITS(6,2)
      COMMON/AQUIFR/HEIGHT,POR,V0,ALPHA,ADSORB,C0,COSUM,VX0,VY0
      COMMON/METRNQ/TITLE(20),NFRNTS,DATE(7),TMAX,DL,DLMIN
      COMMON/DIMENS/NMAX,MXPATH,NSLTOT,NSLARR,NXTMEM,NDIMA
      COMMON A(7000)
C
C     THE ARRAY A IS THE MAIN STORAGE ARRAY FOR RESSQ. THE
C     INDIVIDUAL ARRAYS DIRECTLY USED BY RESSQ ARE STORED AS
C     PART OF THE ARRAY A BY USING EQUIVALENCE STATEMENTS OR
C     SUBROUTINE CALLS. THE FOLLOWING EQUIVALENCE STATEMENTS
C     INTERLEAVE THE ARRAYS XW, QW, RADW, NAMEW AND INDW,
C     CONTAINING INFORMATION ON THE WELLS, IN THE ARRAY A.
C     THE 8 IN THE DIMENSION STATEMENT IS TO SPACE BETWEEN
C     CONSECUTIVE ELEMENTS OF EACH ARRAY TO ALLOW THEM TO BE
C     INTERLEAVED.
C     THE FIRST INDEX FOR XW IS 1 OR 2 FOR THE X OR Y COORDINATE
C     THE FIRST INDEX FOR QW IS 1
C     THE FIRST INDEX FOR RADW IS 1
C     THE FIRST INDEX FOR NAMEW IS 1, 2 OR 3 AS NAME TAKES 3 WORDS
C     THE FIRST INDEX FOR INDW IS 1
C     THE SECOND INDEX FOR EACH ARRAY IDENTIFIES THE WELL
C
      DIMENSION XW(8,1),QW(8,1),RADW(8,1),NAMEW(8,1),INDW(8,1)
      EQUIVALENCE(A(1),XW(1,1))
      EQUIVALENCE(A(3),QW(1,1))
      EQUIVALENCE(A(4),RADW(1,1))
      EQUIVALENCE(A(5),NAMEW(1,1))
      EQUIVALENCE(A(8),INDW(1,1))
C
C     UNIT CONVERSION INFORMATION
C
      DIMENSION COEF(5),COEFS(5,2)
      EQUIVALENCE (COEF,COEFL),(UNIT,UNITL)
```

```
      DATA COEFS/5*1.,100.,277.7778,3.17E-6,3.15576E7,100./
      DATA UNITS/'CM','CM3/S','CM/S','SECONDS',' CM',' CGS',
     1 'METERS','M3/H','M/YR','YEARS','METERS','PRACTICAL'/
C
C     LENGTH OF ARRAY A
C
      DATA NDIMA/7000/
C
C     START READING INPUT
C
      READ(5,5010,END=998) TITLE
C
C     READ NUMBER OF INJECTION WELLS, NUMBER OF PRODUCTION
C     WELLS, AMBIENT AQUIFER CONCENTRATION, DEFAULT INJECTION
C     CONCENTRATION, CONCENTRATION UNITS, FLAG FOR SYSTEM OF
C     UNITS.
C
   10 READ(5,5020,END=998) NWI,NWP,C0,CD,UNITC,NSYST
      IF (NWI.LE.0) GO TO 30
C
C     DECIDE WHICH UNITS ARE TO BE USED FOR I/O (CGS OR PRACTICAL)
C
      IF (UNITC.EQ.' ') UNITC='PERCENT'
      IF (NSYST.NE.1) NSYST=2
      DO 1010 I=1,5
        COEF(I)=COEFS(I,NSYST)
 1010   UNIT(I)=UNITS(I,NSYST)
      UNIT(6)=UNITS(6,NSYST)
C
C     READ AQUIFER THICKNESS, POROSITY, PORE VELOCITY OF REGIONAL
C     FLOW, ANGLE OF REGIONAL FLOW TO +X AXIS, ADSORPTION
C     CAPACITY OF ROCK MATRIX.
C
   20 READ(5,5030,END=999) HEIGHT,POR,V0,ALPHA,ADSORB
C
C     FIND THE STARTING LOCATIONS IN ARRAY A FOR THE ARRAYS
C     NPATH, ITRW, ICW, AND BETA.
C     NWELL   = NUMBER OF WELLS
C     IP1     = INDEX OF FIRST PRODUCTION WELL
C     NPATH1  = INDEX IN ARRAY A OF FIRST ELEMENT OF ARRAY NPATH
C     ITR1    = INDEX IN ARRAY A OF FIRST ELEMENT OF ARRAY ITRW
C     ICI1    = INDEX IN ARRAY A OF FIRST ELEMENT OF ARRAY CIW
C     IBETA1  = INDEX IN ARRAY A OF FIRST ELEMENT OF ARRAY BETA
C     NXTHEM  = NEXT AVAILABLE LOCATION IN ARRAY A
C
      NWELL=NWI+NWP
      IP1=1+NWI
      NPATH1=1+8*NWELL
      ITR1=NPATH1+NWI
      ICI1=ITR1+NWI
      IBETA1=ICI1+NWI
      NXTMEM=IBETA1+NWI
C
C     CONTINUE WITH THE MAIN PROGRAM IN SUBROUTINE SETDIM USING
C     THE NAMES NPATH, ITRW, CIW AND BETA FOR THE RESPECTIVE
C     PORTIONS OF THE ARRAY A.
C
      CALL SETDIM(A(NPATH1),A(ITR1),A(ICI1),A(IBETA1))
      GO TO 10
```

```
 30   IF (NWP.LE.0) STOP
      WRITE(6,6010) NWI,NWP
      STOP
5010  FORMAT(20A4)
5020  FORMAT(2I5,2E10.4,A10,I5)
5030  FORMAT(5E10.4)
6010  FORMAT(13H-THE CASE OF ,I2,15HINJECTION WELLS/5H AND , I2,
     1 51H PRODUCTION WELLS CANNOT BE TREATED BY THIS PROGRAM)
 998  STOP
 999  STOP 'ERROR'
      END
```

```
      SUBROUTINE SETDIM(NPATH,ITRW,CIW,BETA)
      DIMENSION NPATH(1),ITRW(1),CIW(1),BETA(1)
      COMMON/METRNQ/TITLE(20),NFRNTS,DATE(7),TMAX,DL,DLMIN
      COMMON/DIMENS/NMAX,MXPATH,NSLTOT,NSLARR,NXTMEM,NDIMA
      COMMON A(16)
      DIMENSION XW(8,1),QW(8,1),RADW(8,1),NAMEW(8,1),INDW(8,1)
      EQUIVALENCE(A(1),XW(1,1))
      EQUIVALENCE(A(3),QW(1,1))
      EQUIVALENCE(A(4),RADW(1,1))
      EQUIVALENCE(A(5),NAMEW(1,1))
      EQUIVALENCE(A(8),INDW(1,1))
C
C     CHECK TO SEE IF THE ARRAY A IS LARGE ENOUGH TO CONTAIN ALL
C     THE DATA THUS FAR ASSIGNED TO GO INTO IT.  EXTEND THE
C     SIZE OF THE ARRAY A IF POSSIBLE AND INFORM THE USER IF NOT.
C
      CALL EXTEND
      IF (NXTMEM.GT.NDIMA) RETURN
C
C     READ THE INFORMATION ON EACH WELL
C
      CALL READW(NPATH,ITRW,CIW,BETA)
C
C     READ THE NUMBER OF FRONTS, AND THE TIME OF EACH FRONT
C
      READ(5,5040,END=999) NFRNTS,DATE
C
C     FIND THE STARTING LOCATIONS IN ARRAY A FOR ARRAYS TARR,
C     NWARR, XF.
C     NTAR     = INDEX IN A OF FIRST ELEMENT OF ARRAY TARR
C     NPARD    = INDEX IN A OF FIRST ELEMENT OF ARRAY NWARR
C     NXFR     = INDEX IN A OF FIRST ELEMENT OF ARRAY XF
C         (IF NFRNTS = 0 THE ARRAY XF IS NOT NEEDED)
C
      NTAR=NXTMEM
      NPARD=NTAR+NSLTOT
      NXTMEM=NPARD+NSLTOT
      CALL EXTEND
      IF (NXTMEM.GT.NDIMA) RETURN
      NXFR=NXTMEM
      IF (NFRNTS.EQ.0) GO TO 40
```

```
      MXPATH=MXPATH+1
      NXTMEM=NXFR+2*NFRNTS*MXPATH
      CALL EXTEND
      IF (NXTMEM.GT.NDIMA) RETURN
      CALL SORT1(DATE,NFRNTS)
C
C     DIVIDE THE REMAINING SPACE IN THE ARRAY A EVENLY FOR THE
C     ARRAYS XL AND YL, USED FOR STORING THE COORDINATES ON
C     THE STREAMLINE IN SUBROUTINE FLOW, AND TIME AND CONCEN-
C     TRATION IN SUBROUTINE CONCEN. CALCULATE THEIR DIMENSION,
C     NMAX, AND THEIR STARTING LOCATIONS NXL AND NYL. CALL
C     RESSQ2 TO CONTINUE THE MAIN PROGRAM WITH THE ARRAY A COM-
C     PLETELY PARTITIONED. THE DIMENSION OF XL AND YL SHOULD
C     BE AT LEAST 100 FOR USE IN FLOW AND AT LEAST MXPATH FOR
C     USE IN CONCEN.
C
   40 NXL=NXTMEM
      NXTMEM=NXL+MAX0(NDIMA-NXTMEM,2*MXPATH,200)
      CALL EXTEND
      IF (NXTMEM.GT.NDIMA) RETURN
      NMAX=(NDIMA-NXL)/2
      NYL=NXL+NMAX
      NXTMEM=NYL+NMAX
      CALL RESSQ2(NPATH,ITRW,CIW,BETA,A(NTAR),A(NPARD),A(NXFR),
     1 A(NXL),A(NYL),MXPATH)
      RETURN
  999 STOP 'ERROR'
 5040 FORMAT(I5,5X,7E10.4)
      END
```

```
      SUBROUTINE RESSQ2(NPATH,ITRW,CIW,BETA,TARR,NWARR,XF,XL,YL,
     1 NFRDIM)
C
C     THIS IS A CONTINUATION OF THE MAIN PROGRAM BUT WITH
C     SPECIFIC NAMES FOR THE VARIOUS PORTIONS OF THE A ARRAY.
C
      DIMENSION  NPATH(1),ITRW(1),CIW(1),BETA(1),TARR(1),NWARR(1),
     1 XF(2,NFRDIM,1),XL(1),YL(1),NAME(3)
      CHARACTER*8 YMDHMS
      COMMON/WELLS/NWI,NWP,NWELL,IP1,CD
      COMMON/IOUNIT/UNITL,UNITQ,UNITV,UNITT,UNITR,UNISYS,UNITC,
     1           COEFL,COEFQ,COEFV,COEFT,COEFR,NSYST
      CHARACTER*10 UNITL,UNITQ,UNITV,UNITT,UNITR,UNISYS,UNITC
      COMMON/AQUIFR/HEIGHT,POR,V0,ALPHA,ADSORB,C0,COSUM,VX0,VY0
      COMMON/METRNQ/TITLE(20),NFRNTS,DATE(7),TMAX,DL,DLMIN
      COMMON/DIMENS/NMAX,MXPATH,NSLTOT,NSLARR,NXTMEM,NDIMA
      COMMON A(16)
      DIMENSION XW(8,1),QW(8,1),RADW(8,1),NAMEW(8,1),INDW(8,1)
      EQUIVALENCE(A(1),XW(1,1))
      EQUIVALENCE(A(3),QW(1,1))
      EQUIVALENCE(A(4),RADW(1,1))
      EQUIVALENCE(A(5),NAMEW(1,1))
```

```
      EQUIVALENCE(A(8),INDW(1,1))
      TWOPI=8.*ATAN(1.)
C
C     READ PERIOD OF TIME TO STUDY, STEP LENGTH, AND FLAGS--
C       NTL=-1 TO SUPPRESS PLOT OF STREAMLINES
C       NTF=-1 TO SUPPRESS PLOT OF FRONTS
C
   50 READ(5,5050,END=999) TMAX,DL,NTL,NTF
C
C     READ LIMITS OF PLOT AREA, DETERMINE DEFAULT STEP LENGTH.
C
      READ(5,5060,END=999) XMIN,XMAX,YMIN,YMAX
      IF (DL.EQ.0.) DL=(XMAX-XMIN)/200.
C
C     PRINT INPUT
C
      IF (NSYST.EQ.1) WRITE(6,6020) TITLE,UNISYS,V0,UNITV,ALPHA,
     1 HEIGHT,UNITL,POR,TMAX,UNITT,C0,UNITC,CD,UNITC,DL,UNITL,
     2 ADSORB
      IF (NSYST.NE.1) WRITE(6,6030) TITLE,UNISYS,V0,UNITV,ALPHA,
     1 HEIGHT,UNITL,POR,TMAX,UNITT,C0,UNITC,CD,UNITC,DL,UNITL,
     2 ADSORB
      IF (NFRNTS.EQ.0) GO TO 60
      IF (NSYST.EQ.1) WRITE(6,6040) NFRNTS,(DATE(I),UNITT,I=1,NFRNTS)
      IF (NSYST.NE.1) WRITE(6,6050) NFRNTS,(DATE(I),UNITT,I=1,NFRNTS)
      DO 1020 I=1,NFRNTS
 1020   DATE(I)=DATE(I)*COEFT
   60 TMAX=TMAX*COEFT
      IF (TMAX.EQ.0.) TMAX=9.42E08
      DL=DL*COEFL
C
C     PRINT THE INFORMATION ON EACH WELL
C
   70 CALL PRINTW(CIW)
C     WRITE TO THE PLOT FILE--LIMITS OF PLOT AREA,
C     NUMBER OF WELLS, WELL PLOT CHARACTER (*),
C     AND THE WELL LOCATIONS (4 PER LINE).
C
      WRITE(7,7010) XMIN,XMAX,YMIN,YMAX
      IW1=1
      MARK=1H*
      WRITE(7,7020) NWELL,MARK
      DO 1030 I=1,NWELL,4
        L=MIN0(4,NWELL-I+1)
        DO 1040 J=1,L
          XL(J)=XW(1,IW1)/COEFL
          YL(J)=XW(2,IW1)/COEFL
 1040     IW1=IW1+1
        WRITE(7,7030) (XL(J),YL(J),J=1,L)
 1030   CONTINUE
C
      ALFA=ALPHA*TWOPI/360.
      COSA=COS(ALFA)
      SINA=SIN(ALFA)
C
C     SET UP THE CONSTANTS IN THE VELOCITY EQUATION.
C     COSUM  = THE COEFFICIENT OF THE SUMMATION
C     VX0    = THE X CONTRIBUTION OF THE REGIONAL FLOW
```

```
C         VY0       = THE Y CONTRIBUTION OF THE REGIONAL FLOW
C
          HEIGHT=HEIGHT*COEFL
          COSUM=(ADSORB-1.)/(POR*HEIGHT*TWOPI)
          SPEED=(1.-ADSORB)*V0*COEFV
          VY0=SPEED*SINA
          VX0=SPEED*COSA
C
C         FOR EACH OF THE INJECTION WELLS CALCULATE NSL=NPATH(I)
C         STEAMLINES COMING FROM THE WELL AT EQUALLY SPACED ANGLES.
C         EVERY ITR(=ITRW(I))-TH STREAMLINE, STARTING WITH THE FIRST,
C         IS TO BE PLOTTED.  (THE FIRST COMES OUT IN BETA1 DIRECTION).
C
          NSLARR=0
          DO 1050 I=1,NWI
            NSL=NPATH(I)
            IF (NSL.EQ.0) GO TO 1050
            ITR=ITRW(I)
            NCT=MIN0(1,ITR)
            X=XW(1,I)
            Y=XW(2,I)
            BETA1=BETA(I)*TWOPI/360.
            RADIUS=RADW(1,I)
            IF (QW(1,I).EQ.0.) RADIUS=0.
            WRITE(6,6060) (NAMEW(J,I),J=1,3)
C
C         EACH CALL TO SUBROUTINE FLOW TRACES OUT ONE STREAMLINE,
C         WITH COORDINATES STORED IN ARRAYS XL AND YL.
C         NWAR  =  NUMBER OF THE PRODUCTION WELL THE STREAMLINE
                   REACHES, 0 IF NO WELL IS REACHED.
C         TAR   =  TIME OF ARRIVAL AT A WELL OR A STAGNATION POINT,
C                  TIME OF LAST STEP IF TIME RUNS OUT.
C         XF(*,J,K)  =  LOCATION OF J-TH STREAMLINE AT K-TH DATE
C                       (* = 1, 2), FOR PLOTTING FRONTS
C         B          =  ANGLE OF DEPARTURE FROM THE INJECTION WELL
C
            DO 1060 J=1,NSL
              XL(1)=X
              YL(1)=Y
              B=AMOD((J-1)*TWOPI/NSL+BETA1,TWOPI)
              CALL FLOW(NWAR,TAR,RADIUS,B,XL,YL,N,XF,J, NFRDIM)
C
C         FROM TIME OF ARRIVAL, TAR, IN SECONDS, COMPUTE TIME TO
C         PRINT, TIMEP, WITH APPROPRIATE UNITS, YMDHMS.
C
              CALL PTIME(TAR,YMDHMS,TIMEP)
C
C         PRINT THE INFORMATION ON THIS STREAMLINE
C         J = NUMBER OF THE STREAMLINE
C         NAME      = NAME OF WELL REACHED
C         BDEG      = ANGLE B IN DEGREES
C
              NAME(1)=4H+++N
              NAME(2)=4HONE+
              NAME(3)=4H++
              IF (TAR.GE.TMAX) GO TO 80
              NAME(1)=4HSTAG
              NAME(2)=4HNATI
```

```
            NAME(3)=4HON
            IF (NWAR.EQ.0) GO TO 80
            NAME(1)=NAMEW(1,NWAR)
            NAME(2)=NAMEW(2,NWAR)
            NAME(3)=NAMEW(3,NWAR)
   80       BDEG=B*360./TWOPI
            WRITE(6,6070) J,NAME,TIMEP,YMDHMS,BDEG
C
C       IF THE STREAMLINE ARRIVED SAVE THE TIME OF ARRIVAL AND
C       THE INJECTION AND PRODUCTION WELLS
C
            IF (NWAR.EQ.0) GO TO 90
            NSLARR=NSLARR+1
            TARR(NSLARR)=TAR
            NWARR(NSLARR)=NWAR*10000+I
C
C       WRITE THIS STREAMLINE INFORMATION TO THE PLOT FILE ONLY
C       IF STREAMLINES ARE TO BE PLOTTED AND THIS IS AN ITR-TH
C       LINE.  WRITE THE COORDINATES OF EVERY THIRD POINT
C       NTHIRD = TOTAL NUMBER OF POINTS TO BE WRITTEN
C
   90       IF ((NTL.EQ.-1).OR.(J.NE.NCT)) GO TO 1060
            NCT=NCT+ITR
            NTHIRD=(N+2)/3
            WRITE(7,7040) J,(NAMEW(II,I),II=1,3),NAME,NTHIRD
            WRITE(7,7030) (XL(II),YL(II),II=1,N,3)
 1060       CONTINUE
C
C       WRITE THE INFORMATION ON THE FRONTS TO THE PLOT FILE
C       ONLY IF THERE ARE ANY FRONTS, FRONTS ARE TO BE PLOTTED
C       IN GENERAL AND SPECIFICALLY FOR THIS INJECTION WELL, AND
C       THERE ARE MORE THAN FOUR POINTS FROM WHICH THIS FRONT IS
C       TO BE PLOTTED.   ADD ONE MORE POINT JOINING THE FIRST TO
C       THE LAST POINT THUS COMPLETING THE FRONT.
C
          IF ((NFRNTS.LE.0.).OR.(NTF.EQ.-1).OR.(INDW(1,I).EQ.-1).OR.
     1       (NSL.LE.4)) GO TO 1050
          NSL=NSL+1
          DO 1070 J=1,NFRNTS
            XF(1,NSL,J)=XF(1,1,J)
            XF(2,NSL,J)=XF(2,1,J)
            CALL PTIME(DATE(J),YMDHMS,TIMEP)
            WRITE(7,7050) TIMEP,YMDHMS,(NAMEW(II,I),II=1,3),NSL
            WRITE(7,7030) ((XF(II,JJ,J),II=1,2),JJ=1,NSL)
 1070       CONTINUE
 1050     CONTINUE
C
C       SUBROUTINE SORT2 SORTS THE ARRAYS NWARR AND TARR—
C       FIRST BY WELL OF ARRIVAL,
C       THEN BY WELL OF DEPARTURE, AND
C       FINALLY BY TIME OF ARRIVAL.
C       SUBROUTINE CONCEN COMPUTES THE EFFECTS OF THE INJECTION
C       ON EACH PRODUCTION WELL VERSUS TIME.
C
        CALL SORT2(TARR,NWARR,NSLAR)
        CALL CONCEN(CIW,NWARR,TARR,XL,YL,NPATH)
        RETURN
  999   STOP 'ERROR'
 5050   FORMAT(2E10.4,2I5)
```

```
5060    FORMAT(4E10.4)
6020    FORMAT(1H1,15X,5A4/3(16X,5A4/),
        1  3H0  ,A9,24H SYSTEM OF UNITS IS USED/
        2 34H0REGIONAL FLOW, PORE VELOCITY        =,1PE10.3,1X,A8/
        3 34H0ORIENTATION OF REGIONAL FLOW        =,3X,0PF7.2,8H DEGREES/
        4 34H0THICKNESS OF THE AQUIFER            =,1PE10.3,1X,A8/
        5 34H0POROSITY                            =,3X,2PF7.2,8H PERCENT/
        6 34H0PERIOD STUDIED                      =,1PE10.3,1X,A8/
        7 34H0INITIAL AQUIFER CONCENTRATION       =,1PE10.3,1X,A8/
        8 34H0DEFAULT INJECTION CONCENTRATION     =,1PE10.3,1X,A8/
        9 34H0STREAMLINE STEP LENGTH              =,1PE10.3,1X,A8/
        1 34H0ADSORPTION CAPACITY OF ROCK         =,3X,2PF7.2,8H PERCENT)
6030    FORMAT(1H1,15X,5A4/3(16X,5A4/),
        1  3H0  ,A9,24H SYSTEM OF UNITS IS USED/
        2 34H0REGIONAL FLOW, PORE VELOCITY        =,0PF10.2,1X,A8/
        3 34H0ORIENTATION OF REGIONAL FLOW        =,3X,0PF7.2,8H DEGREES/
        4 34H0THICKNESS OF THE AQUIFER            =,0PF10.2,1X,A8/
        5 34H0POROSITY                            =,3X,2PF7.2,8H PERCENT/
        6 34H0PERIOD STUDIED                      =,0PF10.2,1X,A8/
        7 34H0INITIAL AQUIFER CONCENTRATION       =,1PE10.3,1X,A8/
        8 34H0DEFAULT INJECTION CONCENTRATION     =,1PE10.3,1X,A8/
        9 34H0STREAMLINE STEP LENGTH              =,0PF10.2,1X,A8/
        1 34H0ADSORPTION CAPACITY OF ROCK         =,3X,2PF7.2,8H PERCENT)
6040    FORMAT(/1H0,I2,23H FRONTS ARE PLOTTED AT ,3(1PE9.2,1XA8),
        1  2(/10X,4(1PE9.2,1XA8)))
6050    FORMAT(/1H0,I2,22H FRONTS ARE PLOTTED AT,3(1XF6.3,1XA8),
        1  2(/9X,4(1XF6.3,1XA8)))
6060    FORMAT(1H1,42HSTREAMLINES DEPARTING FROM INJECTION WELL ,
        1 3A4//35H   NUMBER OF        WELL          TIME OF,
        2        18H           ANGLE BETA/
        3        35H STREAMLINE    REACHED          ARRIVAL,
        4        18H           IN DEGREES/)
6070    FORMAT(4X,I2,6X,3A4,0PF7.1,1X,A8,4X,0PF5.1)
7010    FORMAT(5H1PAGE/27H0LIMITS OF THE AREA PLOTTED/10X,
        1 6H XMIN=,E10.4,6H XMAX=,E10.4,6H YMIN=,E10.4,6H YMAX=,
        2 E10.4)
7020    FORMAT(38H0POINTS AT WHICH THE WELLS ARE LOCATED/
        1 18H NUMBER OF POINTS=,I4,17H  PLOT CHARACTER=,A1/
        2 4(20H    X         Y      ))
7030    FORMAT((1X,2E9.3,3(2X,2E9.3)))
7040    FORMAT(21H0POINTS ON STREAMLINE,I3,6H FROM ,3A4,
        1  12H — ENDS IN ,3A4/18H NUMBER OF POINTS=,I4/
        2  4(20H    X         Y      ))
7050    FORMAT(19H0LINES TO FORM THE ,F10.4,1X,A8,14H FRONT AROUND
        1 ,3A4/18H NUMBER OF POINTS=,I4/4(20H    X         Y      ))
        END
```

```
      SUBROUTINE CONCEN(CIW,NWARR,TARR,TIME,CONC,NPATH)
C
C     THIS SUBROUTINE STUDIES THE CONCENTRATION AT EACH OF
C     THE PRODUCTION WELLS BASED ON THE ARRIVAL OF
C     STREAMLINES FROM THE INJECTION WELLS
C     CIW         = ARRAY OF INJECTION CONCENTRATIONS
C     NWARR       = ARRAY OF STARTING AND ENDING WELLS FOR THE
C                   STREAMLINES
C     TARR        = ARRAY OF TIMES OF ARRIVAL FOR THE STREAMLINES
C     TIME, CONC  = SCRATCH ARRAYS FOR SAVING TIMES AND
C                   CONCENTRATIONS
C     NPATH       = ARRAY OF NUMBER OF STREAMLINES FROM EACH
C                   WELL (TO KNOW WHAT PORTION OF THE FLOW FROM
C                   EACH WELL IS CARRIED BY EACH STREAMLINE)
C     REASONABLE UNITS ARE CHOSEN FOR THE TIME (BASED ON THE
C     QUICKEST STREAMLINE), AND REASONABLE LIMITS ARE CHOSEN
C     FOR THE CONCENTRTION VS TIME PLOT.
C
      CHARACTER*8 YMDHMS
      COMMON/WELLS/NWI,NWP,NWELL,IP1,CD
      COMMON/IOUNIT/UNITL,UNITQ,UNITV,UNITT,UNITR,UNISYS,UNITC,
     1            COEFL,COEFQ,COEFV,COEFT,COEFR,NSYST
      CHARACTER*10 UNITL,UNITQ,UNITV,UNITT,UNITR,UNISYS,UNITC
      COMMON/AQUIFR/HEIGHT,POR,V0,ALPHA,ADSORB,C0,COSUM,VX0,XY0
      COMMON/DIMENS/NMAX,MXPATH,NSLTOT,NSLARR,NXTMEM,NDIMA
      COMMON A(16)
      DIMENSION XW(8,1),QW(8,1),RADW(8,1),NAMEW(8,1),INDW(8,1)
      EQUIVALENCE(A(1),XW(1,1))
      EQUIVALENCE(A(3),QW(1,1))
      EQUIVALENCE(A(4),RADW(1,1))
      EQUIVALENCE(A(5),NAMEW(1,1))
      EQUIVALENCE(A(8),INDW(1,1))
      DIMENSION NWARR(1),TARR(1),TIME(3),CONC(3),CIW(1),NPATH(1)
      IF ((NWP.LE.0).OR.(NSLARR.LE.1)) RETURN
      JS=0
      DO 1080 I=IP1,NWELL
        IF (INDW(1,I).NE.-1) JS=1
 1080   CONTINUE
      IF (JS.EQ.0) RETURN
C
C     TO DETERMINE THE PLOT LIMITS--
C     FIND THE MINIMUM AND MAXIMUM, CMINI AND CMAXI, OF ALL
C     THE CONCENTRATIONS (C0, CD AND THE INJECTION
C     CONCENTRATIONS), AND EXPAND TO INTEGERS (CC1,CC2).
C
      CMINI=AMIN1(C0,CD)
      CMAXI=AMAX1(C0,CD)
      DO 1090 I=1,NWI
        CC1=CIW(I)
        CMINI=AMIN1(CMINI,CC1)
        CMAXI=AMAX1(CMAXI,CC1)
 1090   CONTINUE
      CC1=INT(CMINI)
      CC2=INT(CMAXI)
      IF (CC1.GT.CMINI) CC1=CC1-1.
      IF (CC2.LT.CMAXI) CC2=CC2+1.
      CDMC0=CD-C0
C     TMN         = TIME OF ARRIVAL OF THE QUICKEST STREAMLINE
C     YMDHMS      = APPROPRIATE UNITS FOR TMN
```

```
C           CNVTIM      = FACTOR NEEDED TO CONVERT SECONDS TO
C                         THOSE UNITS
C           TMX         = TIME OF ARRIVAL OF THE SLOWEST STREAMLINE
C           TMX0        = TMX ROUNDED UP TO A MULTIPLE OF A POWER OF TEN
C
            TMN=TARR(1)
            TMX=TARR(1)
            DO 1100 I=1,NSLARR
              TMN=AMIN1(TMN,TARR(I))
              TMX=AMAX1(TMX,TARR(I))
 1100         CONTINUE
            CALL PTIME(TMN,YMDHMS,TIMEP)
            CNVTIM=TIMEP/TMN
            TMN=TIMEP
            TMX=TMX*CNVTIM
            IEXP2=ALOG10(TMX)+1.
            IF (IEXP2.GE.ALOG10(TMN)+1.)IEXP2=IEXP2-1
            TMX0=10.**IEXP2
            I=TMX/TMX0+.95
            TMX0=I*TMX0
            TMN0=0.
            WRITE(7,7060) TMN0,TMX0,CC1,CC2
C
C           THE LOOP FOR EACH PRODUCTION WELL
C           IFIRST  = INDEX OF FIRST STREAMLINE TERMINATING AT WELL
C           LAST    = INDEX OF LAST STREAMLINE TERMINATING AT WELL
C           NPSUIV = MINIMUM VALUE INDICATING A STREAMLINE
C                     TERMINATING AT A HIGHER NUMBERED WELL
C           INDW    = FLAG TO SUPPRESS STUDY OF CONCENTRATION FOR WELL
C           QPROD   = FLOW RATE FROM THIS WELL
C
            LAST=0
            DO 1110 I=IP1,NWELL
              NPSUIV=(I+1)*10000
              IFIRST=LAST+1
              DO 1120 J=IFIRST,NSLARR
                IF (NWARR(J).GE.NPSUIV) GO TO 100
 1120           LAST=J
  100         IF (INDW(1,I).EQ.-1) GO TO 1110
              IF (LAST.LT.IFIRST) GO TO 1110
              QPROD=ABS(QW(1,I))
C
C           SET UP THE ARRAY OF TIMES AT WHICH TO STUDY THE CONCEN-
C           TRATION AT THIS WELL (TIME 0 AND ALL THE TIMES OF ARRIVAL
C           AT THIS WELL), INITIALIZE THE CONCENTRATION ARRAY AND
C           SORT THE TIMES INTO ORDER
C
C           NPNT   = LENGTH OF THE TIME AND CONC ARRAYS USED
              NPNT=1
              TIME(1)=0.
              CONC(1)=C0
              DO 1130 J=IFIRST,LAST
                NPNT=NPNT+1
                CONC(NPNT)=C0
                TIME(NPNT)=TARR(J)
 1130           CONTINUE
              CALL SORT1(TIME,NPNT)
C
C           REPLACE CLUSTERS OF  TIMES WHICH DIFFER BY LESS THAN 1/10
```

```
C         PERCENT OF THE EARLIEST TIME IN THE CLUSTER BY THE LATEST
C         TIME
C
            NPNT1=NPNT
            NPNT=1
            TOLD=0.
            DO 1140 II=2,NPNT1
               IF (1000*(TIME(II)-TOLD).LT.TOLD) GO TO 1140
               TOLD=TIME(II)
               NPNT=NPNT+1
 1140          TIME(NPNT)=TIME(II)
C
C         ADD THE CONTRIBUTION OF EACH STREAMLINE TERMINATING AT
C         THIS WELL TO THE ARRAY CONC.
C         NDEP      = NUMBERS OF THE WELLS ASSOCIATED TO THE PREVIOUS
C                     STREAMLINE ENTERING THIS WELL
C         NWINJ     = NUMBER OF THE INJECTION WELL FOR THE CURRENT
                      STREAMLINE
C         NPATHS    = TOTAL NUMBER OF STREAMLINES FROM NWINJ
C         QINJ      = FLOW RATE FROM NWINJ
C         CNTRBU    = CONTRIBUTION TO THE PRODUCTION CONCENTRATION
C                     OF A STREAMLINE ORIGINATING AT NWINJ
C
            NDEP=0
            DO 1150 JJ=IFIRST,LAST
               IF (NDEP.EQ.NWARR(JJ)) GO TO 110
               NDEP=NWARR(JJ)
               NWINJ=MOD(NDEP,10000)
               NPATHS=NPATH(NWINJ)
               QINJ=ABS(QW(1,NWINJ))
               CNTRBU=(QINJ*(CIW(NWINJ)-C0))/(QPROD*NPATHS)
  110          CALL INTEGR(TARR(JJ),TIME,CONC,NPNT,CNTRBU)
 1150          CONTINUE
C
C         CONVERT THE TIME INTO THE ABOVE CHOSEN UNITS, YMDHMS,
C         AND PRINT THE TABLE OF EVOLUTION OF CONCENTRATION.
C
            IF (CDMC0.EQ.0.) GO TO 120
            WRITE(6,6080) (NAMEW(J,I),J=1,3),YMDHMS,UNITC
            DO 1160 J=2,NPNT
               DCONC=(CONC(J)-C0)/CDMC0
               TIME(J)=TIME(J)*CNVTIM
               WRITE(6,6090) TIME(J),CONC(J),DCONC
 1160          CONTINUE
            GO TO 130
  120       WRITE(6,6100) (NAMEW(J,I),J=1,3),YMDHMS,UNITC
               DO 1170 J=2,NPNT
                  TIME(J)=TIME(J)*CNVTIM
                  WRITE(6,6090)TIME(J),CONC(J)
 1170             CONTINUE
  130          WRITE(7,7070) (NAMEW(II,I),II=1,3),YMDHMS,NPNT,
           1          (TIME(II),CONC(II),II=1,NPNT)
 1110       CONTINUE
            RETURN
 6080       FORMAT(48H1EVOLUTION OF CONCENTRATION FOR PRODUCTION WELL
           1 ,3A4//45H     TIME IN      CONCENTRATION     (C-C0)/(CD-C0)
           2 /5X,A8,3X,3HIN ,A8/)
 6090       FORMAT(2X,0PF9.3,5X,1PE10.3,9X,0PF6.4)
 6100       FORMAT(48H1EVOLUTION OF CONCENTRATION FOR PRODUCTION WELL
```

```
         1 ,3A4//30H      TIME IN      CONCENTRATION
         2 /5X,A8,3X,3HIN ,A8/)
 7060     FORMAT(5H1PAGE/
         1 42H0LIMITS FOR THE CONCENTRATION VS TIME PLOT/
         2 10X,6H TMIN=,E10.4,6H TMAX=,E10.4,6H CMIN=,E10.4,6H CMAX=,E10.4)
 7070     FORMAT(19H0LINES FOR PLOT OF ,3A4,26H CONCENTRATION VS TIME IN
         1 ,A8/18H NUMBER OF POINTS=,I4,/,4(20H      T          C        )/
         2 (1X,2E9.3,3(2X,2E9.3)))
          END
```

```
          SUBROUTINE EXTEND
C
C         THIS SUBROUTINE EXTENDS THE SIZE OF ARRAY A IF POSSIBLE,
C         AND NECESSARY, TO NXTMEM+1000 WORDS.  NDIMA IS UPDATED.
C         IF THE SIZE CANNOT BE EXTENDED A MESSAGE IS PRINTED.
C
          COMMON/DIMENS/NMAX,MXPATH,NSLTOT,NSLARR,NXTMEM,NDIMA
          COMMON A(1)
          IF (NXTMEM.LE.NDIMA) RETURN
          NDIM=NDIMA
C
C         IF THE LENGTH OF BLANK COMMON CAN BE EXTENDED,
C         INSERT THE NECESSARY CARDS TO DO SO HERE.
C         NDIMA  SHOULD  BE  SET TO  THE  NEW DIMENSION.
C         USING A LARGER DIMENSION FOR ARRAY A (NDIMA) COULD AVOID
C         THE NECESSITY OF EXTENDING BLANK COMMON.
C
          IF (NDIMA.GE.NXTMEM) RETURN
          WRITE(6,6110) NDIM,NDIM,NDIM
          RETURN
 6110     FORMAT(13H-THE SIZE OF ,I5
         1 ,34H WORDS FOR ARRAY A IS INSUFFICIENT
         2 /49H0MODIFY THE FOLLOWING 2 CARDS IN THE MAIN PROGRAM
         3 /17H0          COMMON A(,I5,1H)
         4 /19H           DATA NDIMA/,I5,1H/)
          END
```

```
      SUBROUTINE FLOW(NWAR,TAR,RADIUS,B,XL,YL,N,XF,II,NFRDIM)
C
C     THIS SUBROUTINE FOLLOWS A STREAMLINE FROM AN INJECTION
C     WELL TO A PRODUCTION WELL.  THE FOLLOWING ARE RETURNED-
C     NWAR       = WELL OF ARRIVAL (0 IF IT DOES NOT ARRIVE AT A WELL)
C     TAR        = TIME OF ARRIVAL
C     XL, YL     = THE TRACE OF THE STREAMLINE
C     N          = NUMBER OF POINTS IN ARRAYS XL, YL
C     XF(*,II,K) = LOCATION OF II-TH STREAMLINE AT K-TH DATE,
C                  SAVED IN ORDER TO PLOT THE FRONTS (* = 1, 2)
C     THE FOLLOWING ARE SUPPLIED TO THIS ROUTINE-
C     RADIUS     = RADIUS OF THE INJECTION WELL
C     B          = ANGLE OF DEPARTURE FROM THE INJECTION WELL
C     XL(1),YL(1)= LOCATION OF THE INJECTION WELL (AND FIRST POINT
C                  OF THE STREAMLINE)
C     NFRDIM     = THE (VARIABLE) DIMENSION OF THE ARRAY XF
C                  (= MXPATH)
C
      COMMON/IOUNIT/UNITL,UNITQ,UNITV,UNITT,UNITR,UNISYS,UNITC,
     1            COEFL,COEFQ,COEFV,COEFT,COEFR,NSYST
      CHARACTER*10 UNITL,UNITQ,UNITV,UNITT,UNITR,UNISYS,UNITC
      COMMON/METRNQ/TITLE(20),NFRNTS,DATE(7),TMAX,DL,DLMIN
      COMMON/DIMENS/NMAX,MXPATH,NSLTOT,NSLARR,NXTMEM,NDIMA
      COMMON A(16)
      DIMENSION XW(8,1),QW(8,1),RADW(8,1),NAMEW(8,1),INDW(8,1)
      EQUIVALENCE(A(1),XW(1,1))
      EQUIVALENCE(A(3),QW(1,1))
      EQUIVALENCE(A(4),RADW(1,1))
      EQUIVALENCE(A(5),NAMEW(1,1))
      EQUIVALENCE(A(8),INDW(1,1))
      DIMENSION XL(3),YL(3),XF(2,NFRDIM,1)
      REAL MODV,MODVP
C
C     INITIALIZE AND START THE STREAMLINE.
C     XP,YP = PREVIOUS POSITION ON STREAMLINE
C
      COSB=COS(B)
      SINB=SIN(B)
      XP=XL(1)+RADIUS*COSB
      YP=YL(1)+RADIUS*SINB
      XL(1)=XL(1)/COEFL
      YL(1)=YL(1)/COEFL
      XL(2)=XP/COEFL
      YL(2)=YP/COEFL
C
C     DL1        = CURRENT STEP LENGTH (HALVED IF VELOCITY CHANGES
C                  TOO FAST, DOUBLED ON COMPLETION OF THE STEP TO
C                  A MAXIMUM OF DL)
C     XN,YN      = NEXT (TENTATIVE) POSITION
C     NF         = NUMBER OF FRONTS THUS FAR ENCOUNTERED
C     T          = CURRENT TIME
C     IF THE RADIUS IS 0 THIS IS NOT A REAL INJECTION WELL BUT JUST A
C     POINT FROM WHICH TO TRACE A STREAMLINE.  DL1 IS SET TO 0 IN
C     ORDER TO PREVENT THE STREAMLINE FROM BEING FORCED AWAY
C     FROM THE POINT A DISTANCE OF DL IN THE DIRECTION B.  IT
C     IS RESET TO DL AFTER THE FIRST TIME STEP IS CALCULATED.
      DL1=DL
      DLMIN=DL
      N=2
```

```
      NWAR=0
      IF (RADIUS.EQ.0.) DL1=0.
      XN=XP+DL1*COSB
      YN=YP+DL1*SINB
      NF=0
      T=0.
C
C     CALCULATE VELOCITY VECTOR (VXN,VYN) AT (XN,YN) AND
C     MODULUS, MODV.  IF VELOCITY IS 0 STREAMLINE IS AT A
C     STAGNATION POINT.
C
      CALL VXVY(VXN,VYN,XN,YN)
      MODV= SQRT(VXN*VXN+VYN*VYN)
      IF (MODV.EQ.0.) GO TO 210
C
C     (VX,VY) IS THE VELOCITY VECTOR USED IN THE STEP.
C     TNXTFR  = TIME OF THE NEXT FRONT.
C
      VX=VXN
      VY=VYN
      TNXTFR=DATE(1)
      DT=DL1/MODV
      DL1=DL
C
C     START OF MAIN LOOP
C     COME HERE WHEN (XN,YN) IS ACCEPTED AS THE NEXT POINT
C     ALONG THE STREAMLINE.  ADD (XN,YN) TO THE TRACE AFTER
C     MAKING SURE THAT THERE WILL BE ENOUGH ROOM BY WEEDING
C     OUT EVERY OTHER POINT IF NECESSARY.
C
  140 N=N+1
      IF (N.GT.NMAX) CALL WEED(XL,YL,N)
      XL(N)=XN/COEFL
      YL(N)=YN/COEFL
C
C     SAVE PREVIOUS TIME AS TP AND UPDATE THE TIME T.
C     IF THE STEP WAS HALVED EARLIER, TRY DOUBLING IT FOR THE NEXT
C     STEP, KEEPING IT UNDER THE MAXIMUM STEP OF DL.
C
      TP=T
      T=T+DT
      DLMIN=AMIN1(DLMIN,DL1)
      DL1=AMIN1(DL,2*DL1)
C
C     CHECK TO SEE IF ANY FRONT-TIME WAS PASSED IN THE LAST
C     STEP AND IF SO SAVE CURRENT POINT FOR THE FRONT PLOT.
C
  150 IF ((NFRNTS.LE.NF).OR.(T.LT.TNXTFR)) GO TO 160
      NF=NF+1
      DT=TNXTFR-TP
      XF(1,II,NF)=(XP+VX*DT)/COEFL
      XF(2,II,NF)=(YP+VY*DT)/COEFL
      TNXTFR=DATE(NF+1)
      GO TO 150
C
C     BEGINNING OF CODE TO FIND THE NEXT POINT ALONG THE FRONT.
C     SUBROUTINE TESTAR TESTS WHETHER THIS STREAMLINE HAS
C     ARRIVED AT A PRODUCTION  WELL (I.E. IF THE POINT IS WITHIN
C     LENGTH DL OF THE RADIUS OF A WELL)  NWAR.NE.0 INDICATES
```

```
C         THAT THIS IS SO.  SEE IF THERE IS ANY TIME LEFT.
C
    160   CALL TESTAR(NWAR,DWAR,XWAR,YWAR,RWAR,XN,YN)
          IF (NWAR.NE.0) GO TO 190
          IF (T.GE.TMAX) GO TO 210
C
C         UPDATE  PREVIOUS  POINT LOCATION  AND VELOCITY.
C         (VXP,VYP)     = VELOCITY VECTOR  AT  PREVIOUS  POINT  (XP,YP)
C         MODVP         = ITS MODULUS
C         (VXBP,VYPB)   = ITS UNIT DIRECTION VECTOR
C
          VXP=VXN
          VYP=VYN
          XP=XN
          YP=YN
          MODVP= SQRT(VXP*VXP+VYP*VYP)
          IF (MODVP.EQ.0.) GO TO 210
          VXPB=VXP/MODVP
          VYPB=VYP/MODVP
C
C         FROM THE PREVIOUS POINT (XP,YP) AND ITS VELOCITY VECTOR
C         (VXP,VYP) FIND TENTATIVE POINT (XN,YN) AND ITS VELOCITY
C         VECTOR (VXN,VYN).
C         DT            = TIME STEP ASSOCIATED WITH THE STEP DL1
C         (VXS,VYS)  = SAVED VELOCITY VECTOR (FOR CONVERGENCE TEST)
C
    170   DT=DL1/MODVP
          XN=XP+VXP*DT
          YN=YP+VYP*DT
          VXS=VXP
          VYS=VYP
C
C         IF THE PREVIOUS POINT AND THE TENTATIVE POINT ARE INDIS-
C         TINGUISHABLE, A STAGNATION POINT HAS BEEN APPROXIMATED
          AS NEARLY AS POSSIBLE.
C
          IF (XN.EQ.XP.AND.YN.EQ.YP) GO TO 210
          NITER=0
C
C         IF VELOCITY DIRECTION CHANGE IS TOO GREAT, HALVE THE STEP
C         TAKEN AND TRY AGAIN.
C         (VXPB,VYPB) = UNIT DIRECTION VECTOR AT PREVIOUS POINT
C         (VXNB,VYNB) = UNIT DIRECTION VECTOR AT TENTATIVE POINT
C         (CSQR) = MAGNITUDE SQUARED OF ANGULAR ACCELERATION
C
          CALL VXVY(VXN,VYN,XN,YN)
          MODV= SQRT(VXN*VXN+VYN*VYN)
          If (MODV.EQ.0) GO TO 180
          VXNB=VXN/MODV
          VYNB=VYN/MODV
          AX=VXNB-VXPB
          AY=VYNB-VYPB
          CSQR=AX*AX+AY*AY
          IF (CSQR.LE.1.) GO TO 180
          DL1=DL1/2
          GO TO 170
C
C         AT THIS POINT THE VELOCITIES AT THE PREVIOUS POINT AND AT THE
C         TENTATIVE POINT DO NOT SHOW A DRASTIC CHANGE IN DIRECTION,
```

```
C          AND WE BEGIN THE ITERATION PROCESS TO FIND AN ACCEPTABLE
C          AVERAGE VELOCITY.  THIS IS DONE BY CHECKING THE AVERAGE OF
C          THE VELOCITIES AT THE PREVIOUS AND TENTATIVE POINTS AGAINST
C          THE VELOCITY USED TO CALCULATE THE TENTATIVE POINT.  IF
C          THIS DIFFERS BY MORE THAN ONE PERCENT OF THE MODULUS OF THE
C          ABOVE AVERAGE, THAT AVERAGE IS USED TO COMPUTE A NEW TEN-
C          TATIVE POINT AND THE PROCESS IS REPEATED.
C          (VX,VY)    = AVERAGE VELOCITY VECTOR OF (XY,YP) AND (XN,YN)
C          MODV       = MODULUS OF SAID VECTOR
C          DT         = TIME STEP ASSOCIATED WITH STEP DL1
C          (VXS,VYS)  = VECTOR USED TO COMPUTE TENTATIVE POINT
C          DIF        = RELATIVE DIFFERENCE
C          NITER      = NUMBER OF THE ITERATION
C
  180      VX=(VXN+VXP)*0.5
           VY=(VYN+VYP)*0.5
           MODV= SQRT(VX*VX+VY*VY)
           IF (MODV.EQ.0.) GO TO 210
           DT=DL1/MODV
           XN=XP+VX*DT
           YN=YP+VY*DT
           DVX=VXS-VX
           DVY=VYS-VY
           DIF=SQRT(DVX*DVX+DVY*DVY)/MODV
           VXS=VX
           VYS=VY
           IF (DIF.LE.0.01) GO TO 140
           CALL VXVY(VXN,VYN,XN,YN)
           NITER=NITER+1
C
C          ALLOW 5 ITERATIONS FOR THE VELOCITY VECTOR TO CONVERGE
C          OTHERWISE HALVE THE STEP AND TRY AGAIN.
C
           IF (NITER.LT.5) GO TO 180
           DL1=DL1/2
           GO TO 170
C
C          COME HERE WHEN THE STREAMLINE (XN,YN) HAS ARRIVED WITHIN
C          DISTANCE  DWAR  OF  WELL NWARR (AT (XWAR,YWAR) WITH RADIUS
C          RWAR) AS DETERMINED BY TESTAR.
C          (DX,DY)    = VECTOR FROM WELL TO POINT ON STREAMLINE
C          SINB       = SINE OF THE ANGLE OF ARRIVAL
C          COSB       = COSINE OF THE ANGLE OF ARRIVAL
C          (XN,YN)    = POINT ON THE RADIUS OF THE WELL AT THAT ANGLE
C
  190      DX=XN-XWAR
           DY=YN-YWAR
           IF (DWAR.EQ.0.) GO TO 200
           COSB=DX/DWAR
           SINB=DY/DWAR
           CALL VXVY(VXN,VYN,XN,YN)
           MODV= SQRT(VXN*VXN+VYN*VYN)
           XN=XWAR+RWAR*COSB
           YN=YWAR+RWAR*SINB
           DWAR=DWAR-RWAR
           DT=0.
           IF (MODV.NE.0.) DT=DWAR/MODV
           T=T+DT
C
```

```
C          PLACE THE LAST TWO POINTS OF THE STREAMLINE IN THE ARRAYS
C          XL AND YL DISCARDING 1 OR 2 OF THE PREVIOUS POINTS IF THE
C          ARRAYS ARE FULL.
C
  200      N=MIN0(N+2,NMAX)
           XL(N-1)=XN/COEFL
           YL(N-1)=YN/COEFL
           XL(N)=XWAR/COEFL
           YL(N)=YWAR/COEFL
C
C          WIND UP THIS SUBROUTINE
C          TAR = TIME WHEN THIS STREAMLINE CALCULATION IS STOPPED
C                  (TIME OF ARRIVAL, TIME WHEN A STAGNATION POINT
C                  IS REACHED OR TMAX)
C          IF THERE ARE ANY FRONTS LEFT DRAW THEM THROUGH THE
C          LAST POINT REACHED.
C
  210      TAR=T
           IF (NFRNTS.LE.NF) RETURN
           NF=NF+1
           DO 1180 I=NF,NFRNTS
             XF(1,II,I)=XN/COEFL
             XF(2,II,I)=YN/COEFL
 1180        CONTINUE
           RETURN
           END
```

```
           SUBROUTINE INTEGR(TAR,TIME,CONC,NPNT,CNTRBU)
C
C          THIS SUBROUTINE ADDS THE CONTRIBUTION OF A STREAMLINE TO
C          THE ARRAY CONC FOR TIMES FOLLOWING THE TIME OF ARRIVAL
C          OF THE STREAMLINE.
C
C          TIME      = ARRAY OF TIMES AT WHICH TO CALCULATE THE
C                        CONCENTRATION
C          CONC      = ARRAY OF CONCENTRATIONS CORRESPONDING TO THE
C                        ARRAY TIME
C          NPNT      = LENGTH OF ARRAYS TIME AND CONC
C          CNTRBU    = CONTRIBUTION FROM THIS STREAMLINE
C          TAR       = TIME OF ARRIVAL OF THIS STREAMLINE
C          LOOP BACKWARDS THROUGH THE TIME ARRAY ADDING CNTRBU TO
C          THE ARRAY CONC UNTIL A TIME PREDATING THE TIME OF ARRIVAL
C          IS FOUND
C
           DIMENSION TIME(1),CONC(1)
           DO 1190 K=1,NPNT
             IF (TIME(NPNT-K+1).LT.TAR) RETURN
             CONC(NPNT-K+1)=CNTRBU+CONC(NPNT-K+1)
 1190        CONTINUE
           RETURN
           END
```

```
      SUBROUTINE PRINTW(CIW)
C
C     THIS SUBROUTINE PRINTS THE INFORMATION ON THE WELLS
C     XW(*,I)    = COORDINATES OF THE WELL (* = 1, 2)
C     QW(1,I)    = FLOW RATE INTO/FROM THE WELL
C     RADW(1,I)  = RADIUS OF THE WELL
C     INDW(1,I)  = FLAG TO SUPPRESS--
C                   PLOTTING FRONTS AROUND INJECTION WELLS
C                   STUDYING CONCENTRATION FOR PRODUCTION WELLS
C     CIW(I)     = INJECTION CONCENTRATION
C
      COMMON/DIMENS/NMAX,MXPATH,NSLTOT,NSLARR,NXTMEM,NDIMA
      COMMON/WELLS/NWI,NWP,NWELL,IP1,CD
      COMMON/IOUNIT/UNITL,UNITQ,UNITV,UNITT,UNITR,UNISYS,UNITC,
     1          COEFL,COEFQ,COEFV,COEFT,COEFR,NSYST
      CHARACTER*10 UNITL,UNITQ,UNITV,UNITT,UNITR,UNISYS,UNITC
      COMMON A(16)
      DIMENSION XW(8,1),QW(8,1),RADW(8,1),NAMEW(8,1),INDW(8,1)
      EQUIVALENCE(A(1),XW(1,1))
      EQUIVALENCE(A(3),QW(1,1))
      EQUIVALENCE(A(4),RADW(1,1))
      EQUIVALENCE(A(5),NAMEW(1,1))
      EQUIVALENCE(A(8),INDW(1,1))
      DIMENSION CIW(1)
      IPAGE=1
      IF (NWI.EQ.0) GO TO 220
      IPAGE=MIN0(NWELL/51,1)
      WRITE(6,6120) NWI
      WRITE(6,6130) UNITL,UNITL,UNITQ,UNITC,UNITR
      WRITE(6,6130)
220   DO 1200 I=1,NWELL
        IF (I.EQ.IP1) GO TO 230
        IF (I.GT.IP1) GO TO 240
        Q=-QW(1,I)
        IF (NSYST.EQ.1) WRITE(6,6140) (NAMEW(J,I),J=1,3),
     1          XW(1,I),XW(2,I),Q,CIW(I),RADW(1,I),INDW(1,I)
        IF (NSYST.NE.1) WRITE(6,6150) (NAMEW(J,I),J=1,3),
     1          XW(1,I),XW(2,I),Q,CIW(I),RADW(1,I),INDW(1,I)
        GO TO 250
230     WRITE(6,6160) IPAGE,NWP
        WRITE(6,6130) UNITL,UNITL,UNITQ,UNITR
        WRITE(6,6130)
240     IF (NSYST.EQ.1) WRITE(6,6170) (NAMEW(J,I),J=1,3),
     1          XW(1,I),XW(2,I),QW(1,I),RADW(1,I),INDW(1,I)
        IF (NSYST.NE.1) WRITE(6,6180) (NAMEW(J,I),J=1,3),
     1          XW(1,I),XW(2,I),QW(1,I),RADW(1,I),INDW(1,I)
250     XW(1,I)=XW(1,I)*COEFL
        XW(2,I)=XW(2,I)*COEFL
        QW(1,I)=QW(1,I)*COEFQ
        RADW(1,I)=RADW(1,I)*COEFR
1200    CONTINUE
      RETURN
6120  FORMAT(1H1,28X,I2,16H INJECTION WELLS,
     1 /30H0 WELL NAME          X          Y,
     2 47H         FLOW-RATE CONCENTRATION RADIUS INDICATOR)
6130  FORMAT(14X,5(3X,A8))
6140  FORMAT(1X,3A4,5(2X,1PE9.2),4X,I2)
6150  FORMAT(1X,3A4,3(0PF10.2,1X),2(2X,1PE9.2),4X,I2)
```

```
6160    FORMAT(/I1,27X,I2,17H PRODUCTION WELLS,
       1  /30H0 WELL NAME           X           Y,
       2  37H          FLOW-RATE     RADIUS   INDICATOR)
6170    FORMAT(1X,3A4,4(2X,1PE9.2),5X,I2)
6180    FORMAT(1X,3A4,3(0PF10.2,1X),2X,1PE9.2,5X,I2)
        END
```

```
        SUBROUTINE PTIME(TAR,YMDHMS,TIMEP)
C
C       THIS SUBROUTINE CONVERTS TAR (TIME IN SECONDS) INTO
C       REASONABLE UNITS OF TIME (TIMEP) SUPPLYING THE NAME OF
C       THOSE UNITS IN YMDHMS.
C
        DIMENSION TIMES(5)
        CHARACTER*8 YMDHMS,TUNITS(6)
        DATA TIMES/60.,60.,24.,30.4375,12./
        DATA TUNITS/'SECONDS','MINUTES','HOURS','DAYS','MONTHS',
       1 'YEARS'/
        TIMEP=TAR
C
C       CONTINUE CONVERTING THE TIME INTO COARSER UNITS UNTIL THE
C       NEXT CONVERSION WOULD CAUSE THE TIME TO BE GIVEN IN A
C       FRACTION OF A UNIT
C
        DO 1210 I=1,5
          IF (TIMEP.LT.TIMES(I)) GO TO 260
1210      TIMEP=TIMEP/TIMES(I)
        I=6
 260    YMDHMS=TUNITS(I)
        RETURN
        END
```

```
        SUBROUTINE READW(NPATH,ITRW,CIW,BETA)
C
C       THIS SUBROUTINE READS THE INFORMATION ON THE WELLS. CIW,
C       BETA, NPATH, AND ITRW ARE IGNORED FOR PRODUCTION WELLS.
C       XW(*,I)    = COORDINATES OF THE WELL (* = 1, 2)
C       QW(1,I)    = FLOW RATE INTO/FROM THE WELL (NON-NEGATIVE).
C                    FOR INJECTION WELLS QW=0 INDICATES A REGIONAL-
C                    FLOW-STREAMLINE STARTING POINT.
C       RADW(1,I)  = RADIUS OF THE WELL (DEFAULT 7.5 CM)
C       CIW(I)     = INJECTION CONCENTRATION (DEFAULT CD)
C       BETA(I)    = ANGLE OF FIRST STREAMLINE TO LEAVE WELL
```

```
C         NPATH(I)  = NUMBER OF STREAMLINES FROM THE WELL
C                     (DEFAULT 40)
C         ITRW(I)   = RATIO OF NPATH(I) TO NUMBER OF STREAMLINES
C                     PLOTTED (DEFAULT 4)
C         INDW(1,I) = -1  TO SUPPRESS--
C                     PLOTTING FRONTS AROUND INJECTION WELLS
C                     STUDYING CONCENTRATION FOR PRODUCTION WELLS
C
      COMMON/DIMENS/NMAX,MXPATH,NSLTOT,NSLARR,NXTMEM,NDIMA
      COMMON/WELLS/NWI,NWP,NWELL,IP1,CD
      COMMON/IOUNIT/UNITL,UNITQ,UNITV,UNITT,UNITR,UNISYS,UNITC,
     1          COEFL,COEFQ,COEFV,COEFT,COEFR,NSYST
      CHARACTER*10 UNITL,UNITQ,UNITV,UNITT,UNITR,UNISYS,UNITC
      COMMON A(16)
      DIMENSION XW(8,1),QW(8,1),RADW(8,1),NAMEW(8,1),INDW(8,1)
      EQUIVALENCE(A(1),XW(1,1))
      EQUIVALENCE(A(3),QW(1,1))
      EQUIVALENCE(A(4),RADW(1,1))
      EQUIVALENCE(A(5),NAMEW(1,1))
      EQUIVALENCE(A(8),INDW(1,1))
      DIMENSION NPATH(1),ITRW(1),CIW(1),BETA(1)
      MXPATH=0
      NSLTOT=0
      DO 1220 I=1,NWELL
        READ(5,5070,END=999) (NAMEW(J,I),J=1,3),XW(1,I),XW(2,I),QW(1,I),
     1          RADW(1,I),C,BETA1,NSL,ITR,INDW(1,I)
        IF (RADW(1,I).EQ.0.) RADW(1,I)=7.5/COEFR
        IF (I.GT.NWI) GO TO 1220
        IF (C.EQ.0.) C=CD
        IF (ITR.EQ.0) ITR=4
        IF (NSL.EQ.0) NSL=40
        IF (ITR.EQ.-1) ITR=0
        IF (NSL.EQ.-1) NSL=0
        IF (QW(1,I).EQ.0.) NSL=1
        NSLTOT=NSLTOT+NSL
        IF (INDW(1,I).NE.-1) MXPATH=MAX0(MXPATH,NSL)
        QW(1,I)=-AMAX1(QW(1,I),0.)
        NPATH(I)=NSL
        ITRW(I)=ITR
        CIW(I)=C
        BETA(I)=BETA1
 1220   CONTINUE
      RETURN
  999 STOP 'ERROR'
 5070 FORMAT(2A4,A2,6E10.4,3I2)
      END
```

```
      SUBROUTINE SORT1(TIME,N)
C
C     THIS SUBROUTINE SORTS THE ARRAY TIME INTO ASCENDING ORDER
C     IMAX = INDEX OF THE LAST ELEMENT INVOLVED IN THE SORT
C     JMAX = INDEX OF THE LAST ELEMENT TO CHECK AGAIN
C
      DIMENSION TIME(1)
      JMAX=N
  270 IF (JMAX.LT.2) RETURN
      IMAX=JMAX
      JMAX=0
      DO 1230 J=2,IMAX
        TIME1=TIME(J-1)
        IF (TIME1.LE.TIME(J)) GO TO 1230
        TIME(J-1)=TIME(J)
        TIME(J)=TIME1
        JMAX=J-1
 1230   CONTINUE
      GO TO 270
      END
```

```
      SUBROUTINE SORT2(TARR,NWARR,N)
C
C     THIS SUBROUTINE SORTS THE ARRAYS TARR AND NWARR
C       FIRST BY THE NUMBER OF THE WELL OF ARRIVAL,
C       THEN BY THE NUMBER OF THE WELL OF DEPARTURE,
C       FINALLY BY THE TIME OF ARRIVAL.
C     IMAX = INDEX OF THE LAST ELEMENT INVOLVED IN THE SORT
C     JMAX = INDEX OF THE LAST ELEMENT TO CHECK AGAIN
C
      DIMENSION TARR(1),NWARR(1)
      JMAX=N
  280 IF (JMAX.LT.2) RETURN
      IMAX=JMAX
      JMAX=0
      DO 1240 J=2,IMAX
        NWAR1=NWARR(J-1)
        TAR1=TARR(J-1)
        IF (NWAR1.LT.NWARR(J).OR.
     1         (NWAR1.EQ.NWARR(J).AND.TAR1.LE.TARR(J))) GO TO 1240
        NWARR(J-1)=NWARR(J)
        NWARR(J)=NWAR1
        TARR(J-1)=TARR(J)
        TARR(J)=TAR1
        JMAX=J-1
 1240   CONTINUE
      GO TO 280
      END
```

```
      SUBROUTINE TESTAR(NWAR,DWAR,XWAR,YWAR,RWAR,XN,YN)
C
C     THIS SUBROUTINE TESTS THE POINT XN,YN TO SEE IF IT IS WITHIN
C     DISTANCE DL OF A PRODUCTION WELL.
C     THE FOLLOWING ARE RETURNED BY TESTAR
C    XWAR,YWAR= COORDINATES OF THE PRODUCTION WELL REACHED
C    RWAR      = RADIUS OF THAT WELL
C    NWAR      = NUMBER OF THAT WELL (0 IF THERE IS NO SUCH WELL)
C    DWAR      = DISTANCE TO THAT WELL
C
      COMMON/WELLS/NWI,NWP,NWELL,IP1,CD
      COMMON/METRNQ/TITLE(20),NFRNTS,DATE(7),TMAX,DL,DLMIN
      COMMON A(16)
      DIMENSION XW(8,1),QW(8,1),RADW(8,1),NAMEW(8,1),INDW(8,1)
      EQUIVALENCE(A(1),XW(1,1))
      EQUIVALENCE(A(3),QW(1,1))
      EQUIVALENCE(A(4),RADW(1,1))
      EQUIVALENCE(A(5),NAMEW(1,1))
      EQUIVALENCE(A(8),INDW(1,1))
      IF (NWP.LE.0) RETURN
      DO 1250 NWAR=IP1,NWELL
        IF (QW(1,NWAR).LE.0.) GO TO 1250
        XWAR=XW(1,NWAR)
        YWAR=XW(2,NWAR)
        DX=XN-XWAR
        DY=YN-YWAR
        RWAR=RADW(1,NWAR)
        DWAR=SQRT(DX*DX+DY*DY)
        IF (DWAR.LT.DL+RWAR) RETURN
 1250   CONTINUE
      NWAR=0
      RETURN
      END
```

```
      SUBROUTINE VXVY(VX,VY,X,Y)
C
C     THIS SUBROUTINE COMPUTES THE VELOCITY VECTOR (VX,VY) AT
C     A POINT (X,Y)
C     (DX,DY)    = VECTOR FROM ITH WELL TO (X,Y)
C     DX2DY2     = DISTANCE TO (X,Y) SQUARED
C     COSUM      = COEFFICIENT OF THE SUMMATION FOR POINT
C                  SOURCES/SINKS
C     (VX0,VY0)  = CONTRIBUTION FROM UNIFORM REGIONAL FLOW
C     (VXF,VYF)  = DOUBLET CONTRIBUTION FOR FINITE SOURCES/SINKS
C
      COMMON/WELLS/NWI,NWP,NWELL,IP1,CD
      COMMON/AQUIFR/HEIGHT,POR,V0,ALPHA,ADSORB,C0,COSUM,VX0,VY0
      COMMON A(16)
      DIMENSION XW(8,1),QW(8,1),RADW(8,1),NAMEW(8,1),INDW(8,1)
```

```
      EQUIVALENCE(A(1),XW(1,1))
      EQUIVALENCE(A(3),QW(1,1))
      EQUIVALENCE(A(4),RADW(1,1))
      EQUIVALENCE(A(5),NAMEW(1,1))
      EQUIVALENCE(A(8),INDW(1,1))
      VX=0.
      VY=0.
      VXF=0.
      VYF=0.
      DO 1260 I=1,NWELL
        Q=QW(1,I)
        IF (Q.EQ.0) GO TO 1260
        DX=X-XW(1,I)
        DY=Y-XW(2,I)
        DX2DY2=DX*DX+DY*DY
        IF (DX2DY2.EQ.0) GO TO 290
        VX=VX+Q*DX/DX2DY2
        VY=VY+Q*DY/DX2DY2
        RF=RADW(1,I)*RADW(1,I)/DX2DY2
        IF (RF.GT.1.) GO TO 290
        DOUB=(DX*VX0+DY*VY0)/DX2DY2
        VXF=VXF+RF*(VX0-2.*DX*DOUB)
        VYF=VYF+RF*(VY0-2.*DY*DOUB)
 1260   CONTINUE
      VX=COSUM*VX+VX0-VXF
      VY=COSUM*VY+VY0-VYF
      RETURN
  290 VX=0.
      VY=0.
      RETURN
      END
```

```
      SUBROUTINE WEED(XL,YL,N)
C
C     THIS SUBROUTINE WEEDS OUT EVERY OTHER POINT IN THE TRACE
C     OF THE STREAMLINE STARTING WITH THE FOURTH POINT IN ORDER
C     TO OBTAIN ENOUGH ROOM TO SAVE THE ENTIRE STREAMLINE
C
C     XL, YL = ARRAYS OF COORDINATES OF POINTS
C     N      = NUMBER OF POINTS IN EACH ARRAY
C     N1     = PREVIOUS INDEX OF THE POINTS
C
      DIMENSION XL(1),YL(1)
      N=(N-4)/2+3
      N1=5
      DO 1270 I=4,N
        XL(I)=XL(N1)
        YL(I)=YL(N1)
 1270   N1=N1+2
      N=N+1
      RETURN
      END
```

```
      PROGRAM ZQWELL
      READ (5,1) XREF,YREF,DIST,WIDTH
 1    FORMAT(4E10.4)
      READ (5,2) H,PHI,V0,ALPHA,Q1,N1,NSYST
 2    FORMAT(5E10.4,2I5)
      PI=4.*ATAN(1.)
      COEF=1.
      IF(NSYST.NE.1)COEF=365.25*24.
      ALPHA=2.*PI/360.*ALPHA
      ALPHA1=ALPHA-PI/2.
      D=Q1/(N1*PHI*H*V0)*COEF
      DX=D*COS(ALPHA1)
      DY=D*SIN(ALPHA1)
      NZQW=IFIX(WIDTH/D+.5)+1
      WIDTH=D*FLOAT(NZQW-1)
      XREF=XREF-DIST*COS(ALPHA)-WIDTH/2.*COS(ALPHA1)
      YREF=YREF-DIST*SIN(ALPHA)-WIDTH/2.*SIN(ALPHA1)
      DO 10 K=1,NZQW
        X=XREF+(K-1)*DX
        Y=YREF+(K-1)*DY
        WRITE (6,20) K,X,Y
20      FORMAT(6HZQWELL,I2,2X,2E10.4,3H 0.)
10      CONTINUE
      STOP
      END
```

Appendix I

RT: A Computer Program for Mapping Concentration Distribution in an Aquifer Based on a Time Series Data Collection Concept

The computer program RT converts a time series of concentration data from one or more observation wells into a spatial concentration distribution in the aquifer at various times for cases when regional flow can be neglected and a single production well creates a radial flow field in an aquifer (see section 3.4). A user's guide for the program and a listing of the code are given below.

User's Guide

All input is in list-directed form (free format); items in a given read statement may be on successive lines or on one line separated by commas.

	Variable	*Description*
Read 1		
	XP, YP	Coordinates of production well (m)
Read 2		
	HEIGHT	Aquifer thickness (m)
	PHI	Aquifer porosity (0–1.)
	Q	Production (negative) or injection (positive) flow rate (m^3/h)
	ADSORB	Adsorption capacity of rock matrix, range 0–1; 0 indicates no adsorption, 1 indicates total adsorption
Read 3		
	NT2	Number of spatial concentration distributions to calculate, maximum 5
	(T2(L), L=1, NT2)	Times of spatial concentration distributions (hours)
Read 4		
	NCC	Number of contour levels to plot using the simple contouring scheme described below, maximum 5. If NCC = 0, no contouring is done.

IF NCC>0, Read 5

(CC(K), K=1, NCC)	Concentration contour levels to plot. Data from each well is linearly interpolated or extrapolated to CC values, then segments are drawn connecting equal CC values between adjacent wells. The concentration versus time data from each well must be monotonically increasing (for Q positive) or decreasing (for Q negative) to use this contouring technique.

Read 6

NO	Number of observation wells from which concentration versus time data is obtained, maximum 10

Read 7

NTHETA	Number of angles to assign production well data to, maximum 8. If there is no data from the production well, NTHETA = 0. If NTHETA $\geq$ 0, NO + NTHETA - 1 $\leq$ 10 must be satisfied.

If NTHETA>0 Read 8

(ATHETA(I), I=1, NTHETA)	Angles to assign production well data to (degrees)

Read 9

XMIN, XMAX YMIN, YMAX	Limits of the area studied (m)

Repeat the following NO times

Read 10

XO, YO	Coordinates of observation well (m). If XO=XP and YO=YP (i.e., this observation well is the production well) put it at the end of the list of wells.

Read 11

NTT	Number of time, concentration pairs for this well, maximum 100
(T(J), C(J), J=1, NTT)	Time (hours), concentration data

RT produces two output files. The first, called OUTPUT, lists the parameters of the problem. The second, called TAPE7, is divided into four parts. Each part contains the data needed for a different type of plot; data for each plot begins with the word PAGE.

1. *Time, concentration data.* The following information is given for each

well: time data extrema; concentration data extrema; number of data points; and time, concentration data pairs (four pairs on each line).

2. *Radial distance, concentration data.* The following information is given for each well: radial distance data extrema; concentration data extrema; and for each time T2: number of data points; plot character; and radial distance (from the production well), concentration data pairs (four pairs on a line). A radial distance of -1 indicates that the corresponding concentration has been withdrawn through the production well at time T2.

3. *Spatial concentration distribution.* The following information is given for each time T2: x and y extrema as given by the user in the input file; production well plot character and coordinates (x, y); number of ovservation wells; observation well plot character; coordinates (x, y) of the observation wells; and x, y, concentration data from each well (one x, y, C triplet per line). If the (x, y) coordinates given are the production well coordinates then the corresponding concentration has been withdrawn through the production well.

4. *Simple contouring results.* The following information is given at each time T2: x and y extrema as given by the user in the input file; production well plot character and coördinates (x, y); number of observation wells; observation well plot characters; coordinates (x, y) of the observation wells; and coordinates (x, y) of the end points of segments connecting equal concentrations between adjacent wells, along with a flag for each segment (one segment described per line). The flag has a value from 0 to 3 indicating: (0) the segment connects points interpolated from the data; (1) the segment connects at least one point extrapolated from the data; (2) the segment connects points calculated from at least one data point that has been withdrawn through the production well; and (3) the segment should not be drawn because the contour is not well defined or the angle between the end point wells is too great (>180°).

```
      PROGRAM RT
C                  (INPUT,OUTPUT,TAPE7,TAPE5=INPUT,TAPE6=OUTPUT)
      DIMENSION T(100,10),C(100,10),NT(10),T2(5),XO(10),YO(10), CC(5),
     1 ATHETA(8),R(100,5,10),RO(10),THETAO(10),SIM(5),
     2 RC(5,5,10),IFLAG(5,5,10),THETAOH(10),CH(100,10),RH(100,5,10),
     3 NTH(10),KTH(10),XC(2),YC(2)
      DATA SIM/1H.,1H+,1HO,1HX,1H*/
      PI=4.*ATAN(1.)
C...DIMENSIONS OF ARRAYS
C  NTO  =MAXIMUM NUMBER OF DATA POINTS PER WELL
C  NT2O =MAXIMUM NUMBER OF SPATIAL CONCENTRATION DISTRIBUTIONS
C  NOO  =MAXIMUM NUMBER OF OBSERVATION WELLS
C  NCCO =MAXIMUM NUMBER OF CONTOUR LEVELS TO PLOT
      NTO=100
      NT2O=5
      NOO=10
      NCCO=5
C...READ INPUT
      READ(5,*)XP,YP
      READ(5,*)HEIGHT,PHI,Q,ADSORB
      READ(5,*)NT2,(T2(L),L=1,NT2)
      READ(5,*)NCC
      IF(NCC.GT.0)READ(5,*)(CC(K),K=1,NCC)
      READ(5,*)NO
```

```
      READ(5,*)NTHETA
      IF(NTHETA.GT.0)READ(5,*)(ATHETA(I),I=1,NTHETA)
      READ(5,*)XMIN,XMAX,YMIN,YMAX
      DO 1010 I=1,NO
        READ(5,*)XO(I),YO(I)
        READ(5,*)NTT,(T(J,I),C(J,I),J=1,NTT)
        NT(I)=NTT
 1010   CONTINUE
C...WRITE INPUT
      WRITE(6,6010)
 6010 FORMAT(11H1INPUT DATA,/)
      WRITE(6,6020)XP,YP,HEIGHT,PHI,Q,ADSORB
 6020 FORMAT(28H PRODUCTION WELL COORDINATES/
     1 1X,3HXP=,F7.2,10H M       YP=,F7.2,3H M  /
     2 19H AQUIFER THICKNESS=,F6.2,3H M /
     3 18H AQUIFER POROSITY=,F4.2/
     4 21H PRODUCTION FLOWRATE=,F7.2,7H M3/HR /
     5 21H ADSORPTION CAPACITY=,F4.2)
      WRITE(6,6030)NT2,(T2(L),L=1,NT2)
 6030 FORMAT(44H NUMBER OF SPATIAL DISTRIBUTIONS CALCULATED=,I3/
     1 41H TIMES OF SPATIAL DISTRIBUTIONS IN HOURS=,(5F9.2))
      IF(NCC.GT.0)WRITE(6,6040)(CC(K),K=1,NCC)
 6040 FORMAT(35H SIMPLE CONTOURING—CONTOUR LEVELS=, 1P5E9.2)
      IF(NTHETA.GT.0)WRITE(6,6050)NTHETA,(ATHETA(I),I=1,NTHETA)
 6050 FORMAT(39H NUMBER OF ANGLES PRODUCTION WELL DATA
     1 12H ASSIGNED TO=,I2/16H ANGLES IN DEG.=,8(F5.1,2X))
C...WRITE CONCENTRATION VERSUS TIME DATA TO TAPE7
      DO 1020 I=1,NO
        NTT=NT(I)
        CALL MAXMIN(C(1,I),NTO,1,1,NTT,1,1,CMIN,CMAX)
        CALL MAXMIN(T(1,I),NTO,1,1,NTT,1,1,TMIN,TMAX)
        WRITE(7,7010)I,XO(I),YO(I),TMIN,TMAX,CMIN,CMAX,NTT,(T(J,I),
     1  C(J,I),J=1,NTT)
 7010   FORMAT(42H1PAGE  CONCENTRATION VERSUS TIME FOR WELL ,
     1  I3, 7H  AT X=,F8.2,5H,  Y=,F8.2/15H0LIMITS OF PLOT/10X,
     2  6HTMIN=,E10.4,6H TMAX=,E10.4,6H CMIN=,E10.4, 6H CMAX=,
     3  E10.4/7H0POINTS/18H NUMBER OF POINTS=,I4/
     4  4(20H        T          C       )/(1X,2E9.3,3(2X,2E9.3)))
 1020   CONTINUE
C...CALCULATION OF C(R,T2) FROM C(RO,T)
      A=Q*(1-ADSORB)/(PI*HEIGHT*PHI)
      DO 1030 I=1,NO
        NTT=NT(I)
        DELX=XO(I)-XP
        DELY=YO(I)-YP
        RO(I)=SQRT(DELX*DELX+DELY*DELY)
        IF(RO(I).NE.0.0) GO TO 10
        THETAO(I)=PI/180.*ATHETA(1)
        GO TO 20
   10   THETAO(I)=ATAN2(DELY,DELX)
        IF(THETAO(I).LT.0.)THETAO(I)=THETAO(I)+2.*PI
   20   CONTINUE
        DO 1040 J=1,NTT
          DO 1050 L=1,NT2
            ARG=A*(T2(L)-T(J,I))+RO(I)*RO(I)
            R(J,L,I)=-1.
            IF(ARG.GE.0.)R(J,L,I)=SQRT(ARG)
 1050       CONTINUE
```

```
 1040        CONTINUE
C...WRITE CONCENTRATION VERSUS RADIAL DISTANCE DATA TO TAPE7
          CALL MAXMIN(C(1,I),NTO,1,1,NTT,1,1,CMIN,CMAX)
          CALL MAXMIN(R(1,1,I),NTO,NT2O,1,NTT,NT2,1,RMIN,RMAX)
          IF(RMIN.LT.0.)RMIN=0.
          THETA=180./PI*THETAO(I)
          WRITE(7,7020)I,RO(I),THETA,RMIN,RMAX,CMIN,CMAX
 7020     FORMAT(39H1PAGE  CONCENTRATION VERSUS R FOR WELL ,I3,
     1    6H AT R=,F8.2,8H, THETA=,F6.2,5H DEG./15H0LIMITS OF PLOT/10X,
     2    6HRMIN=,E10.4,6H RMAX=,E10.4,6H CMIN=,E10.4,6H CMAX=,E10.4)
          DO 1060 L=1,NT2
             WRITE(7,7030)T2(L),NTT,SIM(L),(R(J,L,I),C(J,I),J=1,NTT)
 7030        FORMAT(16H0POINTS AT TIME=,E9.3/18H NUMBER OF POINTS=,
     1       I4,17H PLOT CHARACTER=,A1/4(20H        R          C       )/
     2       (1X,2E9.3,3(2X,2E9.3)))
 1060        CONTINUE
 1030     CONTINUE
C...ASSIGN C(R) FOR PROD WELL TO ADDITIONAL ANGLES
       NOT=NO+NTHETA-1
       IF(NOT.LE.NO) GO TO 30
       NOP1=NO+1
       NTT=NT(NO)
       DO 1070 I=NOP1,NOT
         II=I-NO+1
         THETAO(I)=PI/180.*ATHETA(II)
         NT(I)=NT(NO)
         DO 1080 J=1,NTT
            C(J,I)=C(J,NO)
            DO 1090 L=1,NT2
               R(J,L,I)=R(J,L,NO)
 1090          CONTINUE
 1080       CONTINUE
 1070    CONTINUE
       NO=NOT
   30  CONTINUE
C...WRITE X, Y, C DATA TO TAPE7
       DO 1100  L=1,NT2
         WRITE(7,7040)T2(L)
 7040    FORMAT(34H1PAGE  CONCENTRATION DATA AT TIME=,E9.3)
         WRITE (7,7050) XMIN,XMAX,YMIN,YMAX,SIM(3),XP,YP,
     1   NO,SIM(4),(XO(I),YO(I),I=1,NO)
 7050    FORMAT(15H0LIMITS OF PLOT/10X,6H XMIN=,E10.4,6H XMAX=,
     1   E10.4,6H YMIN=, E10.4,6H YMAX=, E10.4/
     2   25H0POINT OF PRODUCTION WELL/
     3   22H NUMBER OF POINTS= 1,17H  PLOT CHARACTER=,A1/
     4   20H      X           Y        /1X,2E9.3/
     5   28H0POINTS OF OBSERVATION WELLS/
     6   18H NUMBER OF POINTS=,I4,
     7   17H  PLOT CHARACTER=,A1/4(20H        X          Y       )/
     8   (1X,2E9.3,3(2X,2E9.3)))
         DO 1110 I=1,NO
            WRITE(7,7060)I,NT(I),SIM(2)
 7050       FORMAT(32H0TRIPLETS--X,Y,C DATA FROM WELL ,I3/
     1      18H NUMBER OF POINTS=,I4,17H PLOT CHARACTER=,A1/
     2      26H     X           Y              C)
            NTT=NT(I)
```

```
                DO 1120 J=1,NTT
                   IF(R(J,L,I).GE.0.) GO TO 40
                   X=XP
                   Y=YP
                   GO TO 50
   40              X=XP+R(J,L,I)*COS(THETAO(I))
                   Y=YP+R(J,L,I)*SIN(THETAO(I))
   50              WRITE(7,7080)X,Y,C(J,I)
 1120              CONTINUE
 1110           CONTINUE
 1100        CONTINUE
          IF(NCC.EQ.0)STOP
C...DO SIMPLE INTERPOLATION
C...ORDER WELLS BY ANGLE THETA
          DO 1130 I=1,NO
 1130        KTH(I)=0
          DO 1140 K=1,NO
             THETMIN=2*PI
             DO 1150 I=1,NO
                IF(THETAO(I).GE.THETMIN.OR.KTH(I).NE.0) GO TO 1150
                THETMIN=THETAO(I)
                IMIN=I
 1150           CONTINUE
             KTH(IMIN)=1
             NTH(K)=NT(IMIN)
             THETAOH(K)=THETAO(IMIN)
             NTT=NT(IMIN)
             DO 1160 J=1,NTT
                CH(J,K)=C(J,IMIN)
                DO 1170 L=1,NT2
                   RH(J,L,K)=R(J,L,IMIN)
 1170              CONTINUE
 1160           CONTINUE
 1140        CONTINUE
C...INTERPOLATE
          DO 1180 I=1,NO
             NTT=NTH(I)
             NTTM1=NTT-1
             DO 1190 K=1,NCC
                DO 1200 L=1,NT2
 1200              IFLAG(K,L,I)=0
C...CHECK FOR EXTRAPOLATION
                IF ( (Q.LT.0.0.AND.CC(K).LE.CH(1,I)) .OR.
     1          (Q.GT.0.0.AND.CC(K).GE.CH(1,I)) ) GO TO 60
                DO 1210 L=1,NT2
 1210              IFLAG(K,L,I)=1
                J=1
                JP1=2
                GO TO 80
   60           IF( (Q.LT.0.0.AND.CC(K).GE.CH(NTT,I) ) .OR.
     1          (Q.GT.0.0.AND.CC(K).LE.CH(NTT,I)) ) GO TO 70
                DO 1220 L=1,NT2
 1220              IFLAG(K,L,I)=1
                J=NTTM1
                JP1=NTT
                GO TO 80
   70           CONTINUE
C...CHECK FOR INTERPOLATION
```

```
                DO 1230 J=1,NTTM1
                   JP1=J+1
                   IF((Q.GT.0..AND.CC(K).GE.CH(J,I).AND.CC(K).LE.CH(JP1,I)).
          1        OR.(Q.LT.0..AND.CC(K).LE.CH(J,I).AND.CC(K).GE.CH(JP1,I)))
          2        GO TO 80
  1230             CONTINUE
    80          CONTINUE
C... DO INTERPOLATION
                DO 1240 L=1,NT2
                   IF(RH(J,L,I).GE.0.)GO TO 90
                   RH(J,L,I)=0.
                   IF(CC(K).NE.CH(JP1,I))IFLAG(K,L,I)=2
    90             CONTINUE
                   IF(RH(JP1,L,I).GE.0.)GO TO 100
                   RH(JP1,L,I)=0.
                   IF(CC(K).NE.CH(J,I))IFLAG(K,L,I)=2
   100             CONTINUE
                   IF(CH(JP1,I).NE.CH(J,I)) GO TO 110
                   RC(K,L,I)=AMAX1(RH(JP1,L,I),RH(J,L,I))
                   IF(IFLAG(K,L,I).EQ.1.OR.IFLAG(K,L,I).EQ.2)IFLAG(K,L,I)=3
                   GO TO 120
   110             RC(K,L,I)=(RH(JP1,L,I)-RH(J,L,I))*(CC(K)-CH(J,I))/
          1        (CH(JP1,I)-CH(J,I))+RH(J,L,I)
                   IF(RC(K,L,I).LT.0.)RC(K,L,I)=0.
   120             CONTINUE
  1240             CONTINUE
  1190          CONTINUE
  1180       CONTINUE
C...WRITE RESULTS OF SIMPLE CONTOURING TO TAPE7
          DO 1250  L=1,NT2
             WRITE(7,7070)T2(L)
  7070       FORMAT(33H1PAGE  SIMPLE CONTOURING AT TIME=,E9.3)
             WRITE(7,7050),XMIN,XMAX,YMIN,YMAX,SIM(3),XP,YP,
          1  NO,SIM(4),(XO(I),YO(I),I=1,NO)
             DO 1260 K=1,NCC
                WRITE(7,7080)CC(K),NO
  7080          FORMAT(40H0PAIRS OF COORDS. CONNECTING CONTOURS OF
          1     ,7H CONC.=,E9.3,24H AND FLAG FOR LINE STYLE/
          2     18H NUMBER OF POINTS=,I4/
          3     47H       X1         Y1         X2         Y2        IPAT)
                DO 1270 I=1,NO
                   IP1=MOD(I,NO)+1
                   XC(1)=RC(K,L,I)*COS(THETAOH(I))
                   YC(1)=RC(K,L,I)*SIN(THETAOH(I))
                   XC(2)=RC(K,L,IP1)*COS(THETAOH(IP1))
                   YC(2)=RC(K,L,IP1)*SIN(THETAOH(IP1))
                   IPAT =0
                   IF(IFLAG(K,L,I).EQ.1.OR.IFLAG(K,L,IP1).EQ.1)IPAT=1
                   IF(IFLAG(K,L,I).EQ.2.OR.IFLAG(K,L,IP1).EQ.2)IPAT=2
                   IF(IFLAG(K,L,I).EQ.3.OR.IFLAG(K,L,IP1).EQ.3)IPAT=3
                   IF(THETAOH(IP1)-THETAOH(I).GE.PI/2.)IPAT=3
C...EACH DATA LINE CONSISTS  OF THE X,Y COORDINATES OF THE
C  ENDPOINTS OF A CONTOUR SEGMENT AND A FLAG IPAT.
C  IPAT  = 0 DRAW A SOLID SEGMENT CONNECTING INTERPOLATED POINTS
C        = 1 DRAW A DASHED SEGMENT CONNECTING EXTRAPOLATED POINT(S)
C        = 2 DRAW A DOTTED SEGMENT CONNECTING POINTS CALCULATED FROM
C            DATA POINT(S) THAT HAVE REACHED THE PRODUCTION WELL.
C        = 3 DO NOT DRAW THIS SEGMENT BECAUSE THE CONTOUR IS NOT WELL
```

```
C          DEFINED OR THE ANGLE BETWEEN THE ENDPOINT WELLS  IS  TOO BIG.
                  WRITE(7,7090)XC(1),YC(1),XC(2),YC(2),IPAT
 7090             FORMAT(2(1X,2E9.3,1X),I5)
 1270             CONTINUE
 1260           CONTINUE
 1250         CONTINUE
          STOP
          END




          SUBROUTINE MAXMIN(A,NO,MO,LO,N,M,L,AMIN,AMAX)
          DIMENSION A(NO,MO,LO),N(LO)
          A11=1.E38
          A22=-A11
          DO 1280 I=1,L
            NN=N(I)
            DO 1290 J=1,M
              DO 1300 K=1,NN
                A11=AMIN1(A11,A(K,J,I))
                A22=AMAX1(A22,A(K,J,I))
 1300           CONTINUE
 1290         CONTINUE
 1280       CONTINUE
          AMIN=INT(A11)
          AMAX=INT(A22)
          IF(AMIN.GT.A11)AMIN=AMIN-1
          IF(AMAX.LT.A22)AMAX=AMAX+1
          RETURN
          END
```

Appendix J

Control of the Movement of a Fluid Plume by Injection and Production Procedures

Introduction

In problems related to contaminant transport, sea water intrusion, and hot water storage in aquifers, it is often very important to control or manipulate the shape and movement of a fluid plume. One means of doing so is to create a fluid flow field around and within the plume through the use of one or more injection or production wells. For example, this flow field may be used to counteract buoyancy flow due to density differences between the plume and native groundwater, thereby maintaining the original shape of the plume and limiting its movement.

One particular problem of interest is the extraction of a contaminant plume from an aquifer. An optimal withdrawal scheme would allow a minimum of uncontaminated groundwater to be removed along with the contaminant. For hot water storage it would maximize the recovery of sensible energy stored in the aquifer. Parameters that greatly influence the movement of a fluid plume during its withdrawal include regional groundwater flow, gravity, and aquifer heterogeneities. These factors need to be considered in the design of an optimal plume withdrawal scheme. The present study addresses this problem by applying a numerical model to study the shape and movement of a hot water plume being withdrawn from an aquifer. These studies are then verified against the results of a recent hot water storage field experiment carried out by Auburn University at Mobile, Alabama.

Methodology

The numerical model PT developed at Lawrence Berkeley Laboratory was used to simulate the movement and deformation of a hot water plume being withdrawn from an aquifer and to calculate various alternative injection and production schemes to optimize the recovery of sensible energy stored in the aquifer. The recovery of energy is measured by the recovery factor ϵ, which is defined as the energy produced divided by the energy injected with energies measured relative to the original aquifer temperature.

PT is an integrated finite difference code that calculates heat and mass transfer in a water–saturated porous or fractured medium. It can be used for one–, two–, or three–dimensional complex geometry problems involving heterogeneous materials. Fluid density is temperature and pressure dependent, and fluid viscosity is temperature dependent. The vertical deformation of the rock matrix may be calculated using the one–dimensional consolidation theory of *Terzaghi* [1925]. Rock thermal conductivity and intrinsic permeability may be temperature dependent and anisotropic. The following physical effects may be included: (1) heat convection and conduction in the aquifer/aquitard system, (2) regional groundwater flow, (3) multiple heat and/or mass sources and sinks, (4) hydrologic or thermal barriers, (5) constant pressure or temperature boundaries, and (6) gravitational effects. PT was developed from the

TABLE J1. Material properties used in the optimal plume withdrawal calculations

Property	Value
Aquifer thermal conductivity	1.5 W/m°C
Rock heat capacity	1000 J/kg°C
Water heat capacity	4057 J/kg°C
Rock density	1200 kg/m^3
Porosity	0.25
Average horizontal permeability	0.63×10^{-10} m^2 (63 darcies)
Individual layer horizontal permeabilities:	
upper layer — 9m thick	0.46×10^{-10} m^2 (46 darcies)
middle layer — 5m thick	1.16×10^{-10} m^2 (116 darcies)
lower layer — 7m thick	0.46×10^{-10} m^2 (46 darcies)
Overall vertical to horizontal permeability ratio	1:7
Storativity	6×10^{-4}

code CCC [*Lippmann et al.*, 1977] which has been used for many years for a variety of energy storage, geothermal, and waste isolation problems. PT employs a much more efficient solution technique than CCC to solve the coupled mass and energy equations. Both PT and CCC have been validated against a large number of analytical solutions and CCC has been used to match the results of several field experiments [*Bodvarsson,* 1982].

Optimal Plume Withdrawal Calculations

A number of withdrawal strategies for a given axisymmetric hot water plume in a confined aquifer were analyzed. In this study we assume that the 21–m–thick aquifer is horizontal and composed of three permeability layers. The middle layer has a transmissivity 2.5 times that of the upper and lower layers. The material properties of the aquifer are summarized in Table J1. The hot plume was created by injecting 80°C water at 0.012 m^3/s (200 gpm) into a well penetrating the entire thickness of the aquifer, which was initially at 20°C. The upper and lower boundaries of the aquifer are insulated no–flow boundaries; there is a constant pressure and temperature boundary at a radial distance of 16 km from the injection well. After an injection period of 40 days, 42,500 m^3 of hot water had been injected. The resulting hot water plume shown in Figure J1 is the starting point for studying the effects of different injection–production schemes on plume extraction. For each scheme the extraction flow rate is 0.012 m^3/s (200 gpm).

From an examination of the shape of the plume (Figure J1) it is clear that a large buoyancy force has caused much of the hot water to rise to the upper part of the aquifer. The high permeability layer has caused a preferential flow into the middle of the aquifer as well. In this axisymmetric problem, regional groundwater flow is not considered and buoyancy flow is the major factor that must be counteracted to obtain a maximum recovery factor. The effect of the aquifer heterogeneity must also be considered.

For contaminant plumes, contaminant concentration C replaces temperature and all the discussions in this paper are still applicable. The different fluid plume withdrawal schemes are summarized in Table J2. Two basic comparisons were made between the

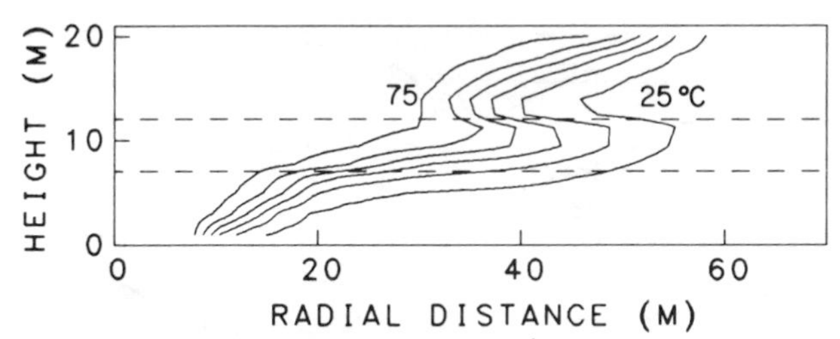

Fig. J1. Calculated temperature distribution in a vertical section of the aquifer showing isotherms of the hot water plume before withdrawal begins. The dashed lines delineate the high-permeability layer.

cases. The first is the recovery factor after 40 days of extraction, when the volume withdrawn equals the original injected plume volume. The second is the extracted volume required to yield a recovery factor of 0.90, when only 10% of the injected energy (or contaminant) is left behind.

Single-Well Withdrawal Schemes

Case 1. Withdrawal through a well that penetrates the full aquifer thickness provides a reference case against which other withdrawal schemes can be compared. For this reference case, calculations using the code PT yield $\epsilon = 0.67$ after 40 days of withdrawal. Examination of the temperature distributions at various times during withdrawal (Figure J2) show that preferential flow out of the high-permeability layer removes the extra heat there quickly and the little heat in the bottom layer of the aquifer is withdrawn quickly as well. The heat in the upper layer of the aquifer is the last to be withdrawn, and much cooler water from the lower layer of the aquifer is withdrawn along with it. A value of $\epsilon = 0.90$ is reached when a volume equal to 1.70 injection volumes has been extracted.

Case 2. Figures J1 and J2 suggest that withdrawal through a well that penetrates only the upper portion of the aquifer may yield a higher value of ϵ at 40 days and a smaller extraction volume for $\epsilon = 0.90$ by withdrawing less of the cool water in the

TABLE J2. Fluid plume withdrawal schemes

Case	PENETRATION INTERVAL (%) Production Well	Auxiliary Injection Wells	ϵ*	Extracted Volume Factor†
1	100	—	0.67	1.70
2	upper 50	—	0.74	1.50
3	upper 40	—	0.77	1.34
4	upper 20	—	0.79	1.27
5	100	100	0.65	1.87
6	100	upper 20	0.68	1.59
7	upper 20	lower 20	0.77	1.37
8	upper 20	upper 20	0.80	1.23

*Recovery factor for extraction volume equal to injected plume volume.

†Extracted volume in units of injected plume volume at $\epsilon = 0.90$, when 10% of the plume is left behind.

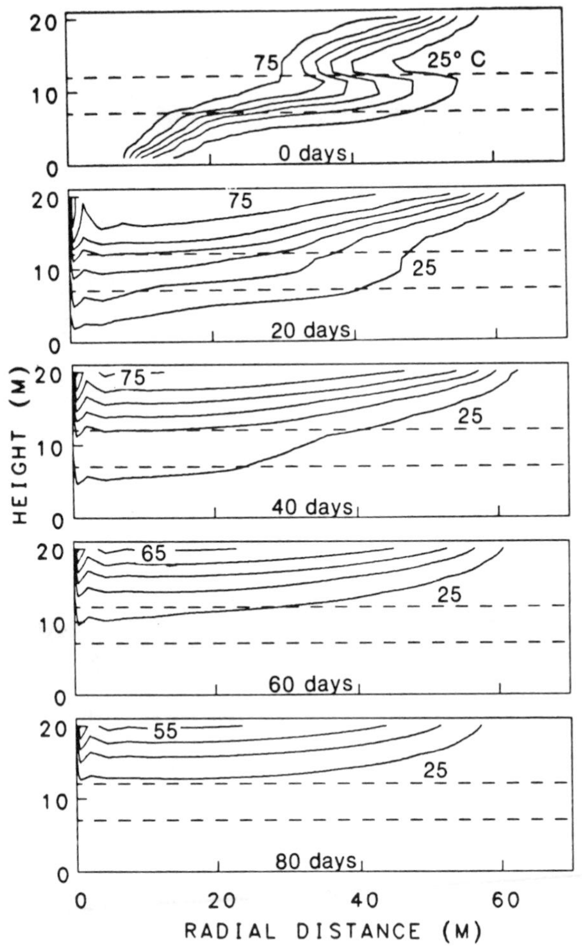

Fig. J2. Calculated temperature distributions during plume withdrawal through a fully penetrating well (case 1).

lower layer of the aquifer. In this case a well penetrating half of the aquifer thickness was used for withdrawal. This scheme gives $\epsilon = 0.74$ at 40 days, a substantial increase over the reference case. The temperature distribution after 40 days of withdrawal (Figure J3*a*) shows that the residual hot fluid remains largely near the top of the aquifer, as in case 1, but that none of the well interval is extracting ambient temperature water at this time, as case 1 did. A value of $\epsilon = 0.90$ is reached when 1.50 injection volumes have been extracted.

Case 3. A case limiting the well penetration to the upper 40% of the aquifer thickness yields $\epsilon = 0.77$ at 40 days, a further increase over the reference case and over case 2. Although the difference in penetration interval between case 2 (10.5 m) and case 3 (8 m) is small, in case 3 the well screen is not open to the high-permeability layer, thus encouraging more fluid to be withdrawn from the upper hotter region of the aquifer. The temperature distribution after 40 days of withdrawal (Figure J3*b*) shows that while most of the residual hot water is near the top of the aquifer, more is

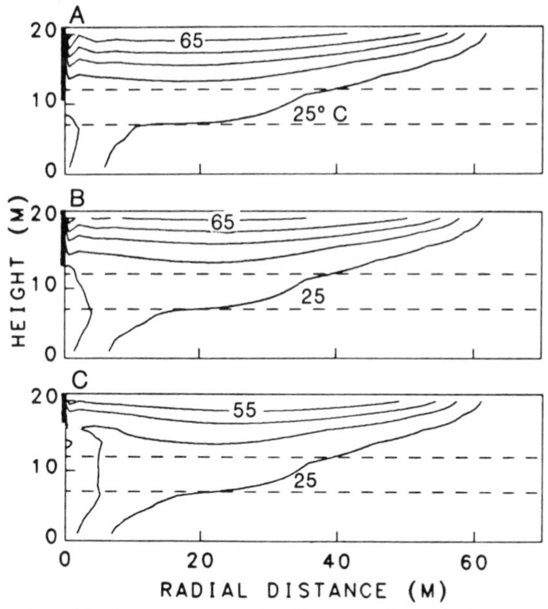

Fig. J3. Calculated residual temperature distribution after 40 days of plume withdrawal through wells penetrating (*a*) the upper half (case 2), (*b*) the upper 40% (case 3), and (*c*) the upper 20% (case 4) of the aquifer thickness. The well screen interval is as shown.

left in the middle and lower aquifer layers than in the previous cases. A value of ϵ = 0.90 is reached when 1.34 injection volumes have been extracted.

Case 4. A case limiting the well penetration to the upper 20% of the aquifer thickness yields ϵ = 0.79 at 40 days. The temperature distribution after 40 days of withdrawal (Figure J3*c*) shows a continuation of the trend seen in cases 2 and 3, better selective withdrawal of water from the upper portion of the aquifer. A value of ϵ = 0.90 is reached when 1.27 injection volumes have been extracted.

Multiple-Well Withdrawal Schemes

Withdrawal schemes using a production well coupled with auxiliary injection wells to modify plume flow have also been studied. In each case a ring of injection wells at a radial distance of r = 62 m, beyond the farthest extent of the plume, is used. These auxiliary wells inject ambient temperature water with a total flow rate equal to the extraction flow rate.

Case 5. This case considers a fully penetrating production well for plume withdrawal and an auxiliary ring of fully penetrating injection wells. This strategy is designed to create a radial flow field during extraction to discourage further buoyancy flow and force the hot water back to the production well. This scheme gives ϵ = 0.65 at 40 days, a decrease from the reference case. Apparently a radial flow field is created, but rather than increasing ϵ by discouraging further buoyancy flow, it forces equal withdrawal of hot and cool water, whereas case 1 selectively produced hot water due to its lower viscosity. Figure J4*a* shows the temperature distribution after 40 days. While the overall temperature distribution is similar to the reference case, the heated region is slightly more compact in this case. A value of ϵ = 0.90 is reached when 1.87 injection volumes have been produced.

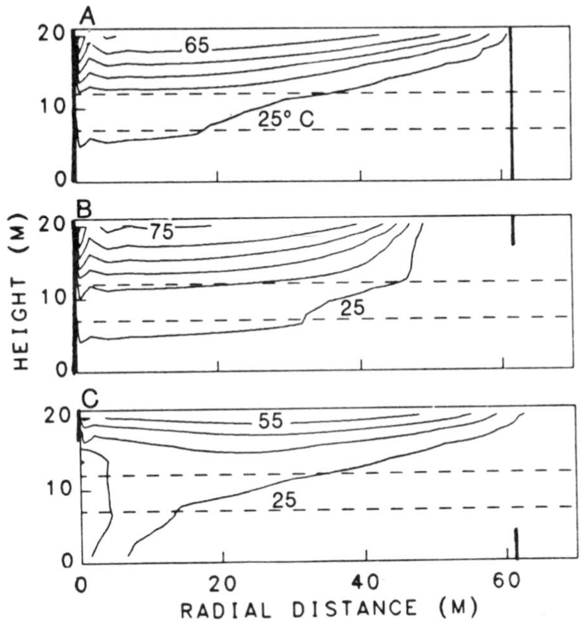

Fig. J4. Calculated temperature distribution after 40 days of plume withdrawal for (*a*) fully penetrating production and auxiliary injection wells (case 5), (*b*) fully penetrating production well and upper 20% auxiliary injection wells (case 6), and (*c*) upper 20% production well and lower 20% auxiliary injection wells (case 7).

Case 6. In order to successfully retard or reverse buoyancy flow, an auxiliary ring of wells that only penetrate the upper 20% of the aquifer thickness is used in conjunction with the fully penetrating production well. This strategy is designed to encourage preferential extraction of the water in the upper layer of the aquifer. This case yields $\epsilon = 0.68$ at 40 days, a modest increase over the reference case. The temperature distribution after 40 days of plume withdrawal (Figure J4*b*) clearly shows the effect the auxiliary ring of wells have on compressing the hot region in the upper layer of the aquifer toward the production well. However, at 40 days ambient temperature water is being produced from the lower layer of the aquifer, as in cases 1 and 5. A value of $\epsilon = 0.90$ is reached after 1.59 injection volumes have been extracted.

Case 7. Of the cases using only a single production well (cases 1–4), an upper 20% penetration (case 4) yields the highest recovery factor at 40 days. In order to examine the effect of coupling auxiliary injection wells with this production interval, two additional cases were studied. Case 7 considers a production well penetrating the upper 20% of the aquifer thickness and an auxiliary ring of injection wells with an open interval in the lower 20% of the aquifer thickness. The case yields $\epsilon = 0.77$ at 40 days, a slight decrease from case 4, which used the upper 20% production well alone. Figure J4*c* shows the temperature distribution after 40 days of plume withdrawal. The auxiliary wells have increased flow into the production well from the high-permeability and lower layers of the aquifer, thus lowering the recovery factor at 40 days. A value of $\epsilon = 0.90$ is reached after 1.37 injection volumes have been extracted.

Case 8. This case considers a production well and an auxiliary ring of injection wells both penetrating the upper 20% of the aquifer thickness. This strategy is

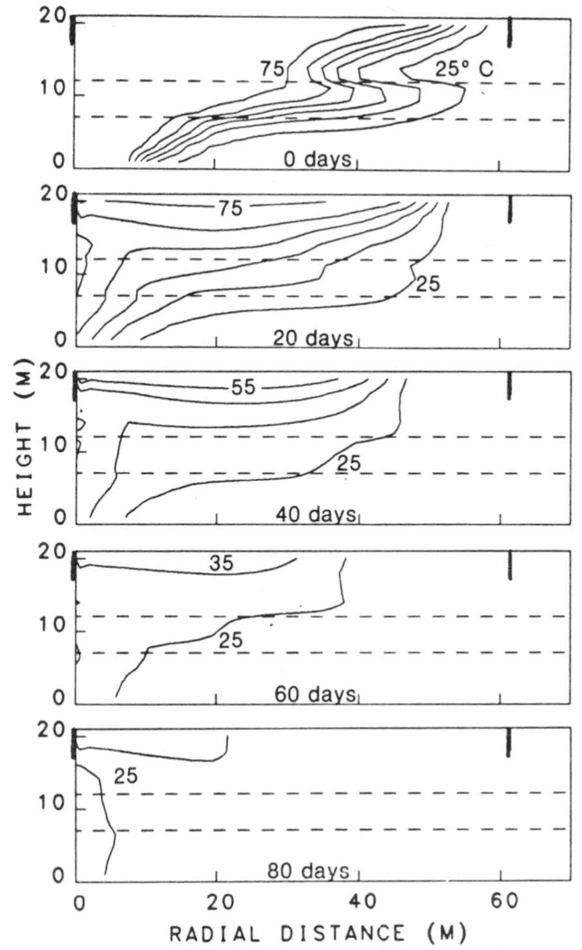

Fig. J5. Calculated temperature distributions during plume withdrawal through an upper 20% production well with upper 20% auxiliary injection wells (case 8).

designed to create a radial flow field in the upper layer of the aquifer, where much of the hot water resides at the beginning of the withdrawal period. This case yields a recovery factor of 0.80, slightly better than case 4, which used the upper 20% production well alone. Figure J5 shows the temperature distribution at various times during the plume withdrawal. The hot region in the upper layer of the aquifer is effectively compressed towards the production well. The high permeability of the middle layer ensures that the heat there is withdrawn, even without a direct connection to the injection and production wells. A value of $\epsilon = 0.90$ is reached when 1.23 injection volumes have been withdrawn.

Based on our two basic comparisons—recovery factor when equal volumes have been injected and produced, and extraction volume required for $\epsilon = 0.90$—case 8 is found to be the optimal withdrawal strategy. However, each of the cases produces particular characteristics that may be desirable for certain applications.

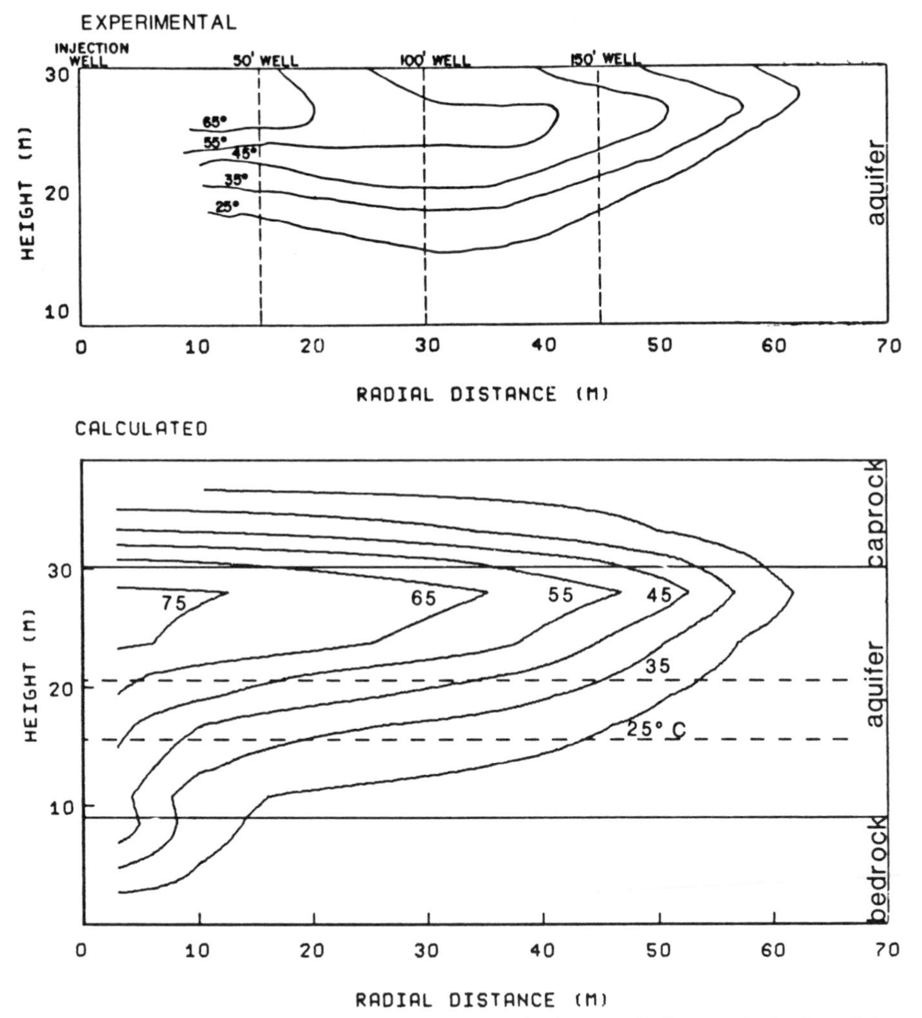

Fig. J6. Experimental and calculated temperature distributions before the beginning of the production period during a hot water storage field experiment.

Verification of Numerical Calculations

During 1981 and 1982 Auburn University conducted an aquifer thermal energy storage field experiment in a shallow aquifer near Mobile, Alabama [*Molz et al.*, 1983]. The second cycle of the experiment involved the injection, storage, and production of 58,000 m^3 of water at an average temperature of 82°C over a 6 month period. The experimental temperature distribution at the end of the storage period, shown in Figure J6, compares well with the calculated temperature distribution obtained with numerical code PT based on actual operating conditions. For the numerical calculation, cap and bedrocks were included in the model because a substantial amount of heat is conducted into the caprock during the injection period. The

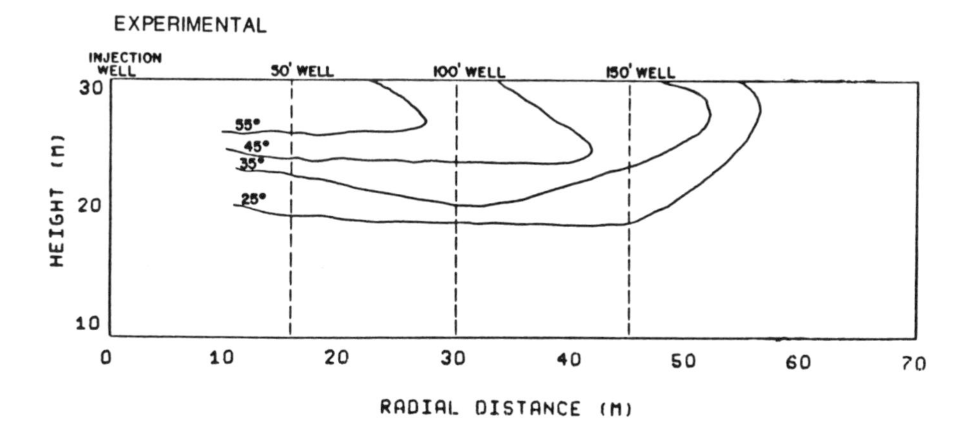

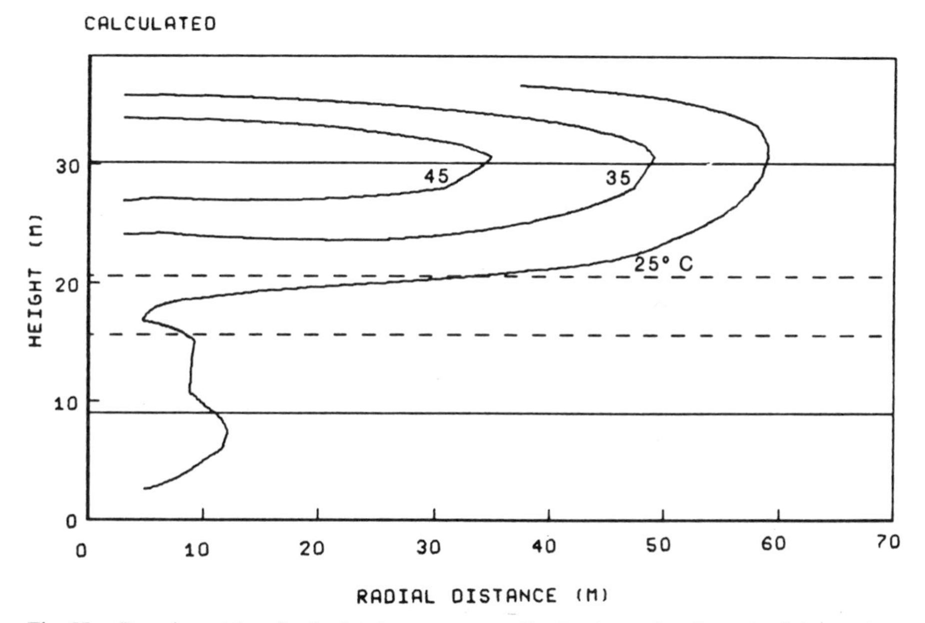

Fig. J7. Experimental and calculated temperature distributions after the end of the production period. The production well initially penetrated the entire aquifer thickness and was then modified to penetrate only the upper half of the aquifer.

material properties used in this calculation are summarized in Table J3.

The withdrawal period was begun using a fully penetrating production well. However, after two weeks of production the well was shut in and modified to produce from only the upper half of the aquifer; then production was resumed. This scenario was numerically simulated. Figure J7 shows the experimental and calculated temperature distributions at the end of the recovery period, indicating quite a good agreement. The experimental recovery factor is 0.452, the calculated one is 0.422 [*Buscheck et al.*, 1983]. This excellent match gives us confidence that the numerical model can calculate physical processes in our study correctly.

TABLE J3. Parameters used in the Auburn field experiment

Parameter	Location	Value
Thermal conductivity	Aquifer	2.29 W/m °C
	Aquitard	2.56 W/m °C
Heat capacity of rock		1.81×10^6 J/m^3°C
Aquifer horizontal permeability		
	Upper layer, 9.6 m thick	0.46×10^{-10} m^2 (46 darcies)
	Middle layer, 5 m thick	1.16×10^{-10} m^2 (116 darcies)
	Lower layer, 6.6 m thick	0.46×10^{-10} m^2 (46 darcies)
	Average value	0.63×10^{-10} m^2 (63 darcies)
Overall vertical to horizontal permeability ratio		1:7
Aquitard to aquifer permeability ratio		10^{-5}
Porosity	Aquifer	0.25
	Aquitard	0.35
Storativity	Aquifer	6×10^{-4}
	Aquitard	9×10^{-2}

Conclusion

In this coupled theoretical study and field data analysis, we have demonstrated the possibility of controlling the movement and shape of a hot water plume during withdrawal using various injection–production arrangements. The results presented here are generally applicable to fluid plumes having different chemical and physical properties than the native groundwater. Further studies are underway to confirm the results presented and to determine appropriate controlling schemes for different scenarios.

Notation

A area, L^2.
$\mathrm{Ai}(x)$ Airy function of x.
a half length of a strip source, L.
b aquifer thickness, L.
C solute concentration, M/L^3.
$\overline{C}$ adsorbed concentration, M/L^3.
C' concentration in a source or sink fluid, M/L^3.
C_D dimensionless concentration.
D hydrodynamic dispersion coefficient, L^2/T.
D_{ij} or D dispersion coefficient tensor, L^2/T.
D_L,D_T longitudinal and transverse dispersion coefficients, L^2/T.
D_x, D_y dispersion coefficient in the x and y directions, L^2/T.
D^* molecular diffusion coefficient, L^2/T.
$\overline{D}$ effective molecular diffusion coefficient, L^2/T.
$\mathrm{erf}(x)$ error function of x, equal to $(2/\sqrt{\pi})\int_0^x e^{-z^2}dz$.
$\mathrm{erfc}(x)$ complementary error function, equal to $1 - \mathrm{erf}(x)$.
$\exp(x)$ exponential of x, equal to e^x.
F mass flux of solute, M/T.
g gravitational acceleration, L/T^2.
H_0 constant head at pond, L.
h hydraulic head, L.
K hydraulic conductivity, L/T.
K_{ij} or K hydraulic conductivity tensor, L/T.
K_d distribution coefficient relating C and $\overline{C}$.
n effective porosity.
n_i directional cosine.
Q rate of recharge or discharge, L^3/T.
Q_p rate of outflow from a pond, L^3/T.
q specific discharge or Darcy velocity, equal to Q/A; L/T.
R_k rate of solute production in reaction k, M/L^3T.
R retardation factor.
r radial distance, L.
r_0 pond radius, L.
r_D dimensionless radius.
r_{Dw} dimensionless well radius.
s parameter of Laplace transformation, $1/T$.
S_s specific storage, $1/L$.
t time, T.
t_D dimensionless time.
t_o period of activity of a source, T.
v average pore water velocity or seepage velocity, equal to Q/nA; L/T.
$\mathbf{v}$ vector of average pore water velocity, L/T.
v_i average pore water velocity in the direction i, L/T.
v_c contaminant velocity, L/T.
v_l velocity of particle along a flow line, L/T.

W	complex velocity potential.
W^*	volume flow rate per unit volume of a source or sink, $1/T$.
x	x coordinate, L.
x_i	Cartesian coordinate, L.
y	y coordinate, L.
α	decay factor of a source, $1/T$.
α_L	longitudinal dispersivity, L.
α_T	transverse dispersivity, L.
λ	radioactive decay constant, equal to ℓn 2/half-life; $1/T$.
ρ_b	bulk density of solid, M/L^3.
ϕ	velocity potential, L^2/T.
ψ	streamline function, L^2/T.
∇	gradient operator, $1/L$.
$\nabla\cdot$	divergence operator, $1/L$.
Δ	difference.

References

Abramowitz, M., and I. A. Stegun, *Handbook of Mathematical Functions*, *Appl. Math. Ser.*, vol. 55, 1046 pp., National Bureau of Standards, Gaithersburg, Md., 1964.

Ahlstrom, S. W., H. P. Foote, R. C. Arnett, C. R. Cole, and R. J. Serne, Multicomponent mass transport model: Theory and numerical implementation (discrete-parcel-random-walk version), *Rep. BNWL-2127*, Battelle Pacific Northwest Lab., Richland, Wash., 1977.

Anderson, D., Does landfill leachate make clay liners more permeable? *Civ. Eng.*, *52*(9), 66-69, 1982.

Anderson, M. P., Using models to simulate the movement of contaminants through groundwater systems, *CRC Crit. Rev. Environ. Control*, *9*(2), 97-156, 1979.

Apgar, M. A., and W. B. Satherwaite, Jr., Groundwater contamination associated with the Llangollen landfill, New Castle County, Delaware: Proceedings of Research Symposium on Gas and Leachate from Landfills, New Brunswick, N.J., report, U.S. Environ. Prot. Agen., Natl. Environ. Res. Center, Cincinnati, Ohio, 1975.

Baca, R. G., R. C. Arnett, and I. P. King, Numerical modeling of flow and transport in a fractured-porous rock system, *Rep. RHO-BWI-SA-113*, Rockwell Hanford Operations, Richland, Wash., 1981.

Bachmat, Y., and J. Bear, The general equations of hydrodynamic dispersion in homogeneous, isotropic porous mediums, *J. Geophys. Res.*, *69*(12), 2561-2567, 1964.

Bachmat, Y., B. Andrews, D. Holtz, and S. Sebastian, Utilization of Numerical Groundwater Models for Water Resource Management, *U.S. Environ. Prot. Agen. Off. Res. Dev. Rep. EPA 600/8-78-012*, 178 pp., 1978.

Bear, J., *Dynamics of Fluids in Porous Media*, 764 pp., American Elsevier, New York, 1972.

Bear, J., *Hydraulics of Groundwater*, 569 pp., McGraw-Hill, New York, 1979.

Bodvarsson, G. S., Mathematical modeling of the behavior of geothermal systems under exploitation, Ph.D. thesis, 353 pp., Univ. of Calif., Berkeley, 1982. (Also available as *Rep. LBL-13937*, Lawrence Berkeley Lab., Berkeley, Calif., 1982.)

Buscheck, T. A., C. Doughty, and C. F. Tsang, Prediction and analysis of a field experiment on a multi-layered aquifer thermal energy storage system with strong buoyancy flow, *Water Resour. Res.*, *19*(5), 1307-1316, 1983.

Campbell, J. E., D. E. Longsine, and R. M. Cranwell, Risk methodology for geologic disposal of radioactive waste: the NWFT/DVM computer code user's manual, report, Sandia Natl. Lab., Albuquerque, N. M., 1981. (Also available as *U.S. Nucl. Regul. Comm. Rep. NUREG/CR-2081*, 1981.)

Claiborne, H. C., L. D. Rickertsen, and R. F. Graham, Expected environments in high level nuclear waste and spent fuel repositories in salt, *Rep. ORNL/TM-7201*, Oak Ridge Natl. Lab., Oak Ridge, Tenn., 1980.

Cleary, R. W., and M. J. Ungs, Groundwater pollution and hydrology, mathematical models and computer programs, *Rep. 78-WR-15*, Water Resour. Program, Princeton Univ., Princeton, N. J., 1978.

Dahlquist, G., and A. Bjorck, *Numerical Methods*, 573 pp., Prentice-Hall, Englewood Cliffs, N. J., 1974.

Davis, S. N., and R. J. M. DeWiest, *Hydrogeology*, 463 pp., John Wiley, New York, 1966.

De Josselin de Jong, G., The tensor character of the dispersion coefficient in anisotropic porous media, in *Fundamentals of Transport Phenomena in Porous Media*, pp. 259-267, Elsevier, New York, 1972.

de Marsily, G., Influence of the spatial distribution of velocities in porous media on the form of solute transport, in Proceedings: Symposium on Unsaturated Flow and Transport Modeling, pp. 299-315, *Rep. PNL-SA-10325*, Pacific Northwest Lab., Richland, Wash., 1982. (Also available as *U.S. Nucl. Regul. Comm. Rep. NUREG/CP-0030*, 1982.)

Dillon, R. T., R. B. Lantz, and S. B. Pahwa, Risk methodology for geologic disposal of radioactive waste: The Sandia waste isolation flow and transport (SWIFT) model, *Rep. SAND78-1267*, Sandia Natl. Lab., Albuquerque, N. M., 1978. (Also available as *U.S. Nucl. Regul. Comm. Rep. NUREG/CR-0424*, 1978.)

Duguid, J. O., and M. Reeves, Dissolved Constituent Transport Code, *IGWMC Key No. 2590*, International Ground Water Modeling Center Database, Holcomb Res. Inst., Indianapolis, Ind., 1976.

Dutt, G. R., M. J. Shafter, and W. J. Moore, Computer simulation model of dynamic physico-chemical processes in soils, *Tech. Bull. 196*, Soil Water Eng. Agric. Exp. Station, Univ. of Arizona, Tucson, 1972.

Edwards, A. L., TRUMP: A computer program for transient and steady state temperature distributions in multidimensional systems, *Rep. UCRL-24754*, Lawrence Livermore Lab., Livermore, Calif., 1972.

Farb, D. G., Upgrading hazardous waste disposal sites, *Rep. SW-677*, U.S. Govt. Printing Off., Washington, D. C, 1978.

Frind, E. O., and P. J. Trudeau, Finite-element analysis of salt transport (SALTRP), *User's Manual Gt-034*, Geotech. Div., Stone and Webster Eng., Inc., Boston, Mass., 1980.

Gringarten, A. C., and J. P. Sauty, Simulation des transferts de chaleur dans les aquiferes, *Bull. Bur. Rech. Geol. Min. Fr., 2nd Series, Sect. III, 1*, 25-34, 1975*a*.

Gringarten, A. C., and J. P. Sauty, A theoretical study of heat extraction from aquifers with uniform regional flow, *J. Geophys. Res.*, *80*(35), 4956-4962, 1975*b*.

Grove, D. B., Ion exchange reactions important in groundwater quality models, in *Advances in Groundwater Hydrology*: Proceedings of a Symposium, Chicago, Ill., pp. 144-152, American Water Res. Assoc., Minneapolis, Minn., 1976.

Grove, D. B., The use of Galerkin finite element methods to solve mass transport equations, *Water Resour. Invest. 77-49*, 55 pp., U.S. Geol. Surv., Denver, Colo., (also available as *PB-277 532*, Natl. Tech. Info. Service, Springfield, Virg., 1977.

Gupta, S. K., C. R. Cole, C. T. Kincaid, and F. E. Kaszeta, Description and application of the FE3DGW and CFEST three-dimensional finite-element models, in Proceedings: Workshop on Numerical Modeling of Thermohydrological Flow in Fractured Rock Masses, *Rep. LBL-11566*, Lawrence Berkeley Lab., Berkeley, Calif., 1980.

Hoopes, J. A., and D. R. F. Harleman, Dispersion in radial flow from a recharging well, *J. Geophys. Res.*, *72*(14), 3595-3607, 1967.

Huyakron, P. S., B. H. Lester, and J. W. Mercer, An efficient finite element technique for modeling transport in fractured porous media, 1, Single species transport, *Water Resour. Res.*, *19*(3), 841-854, 1983.

Intera Environmental Consultants, Hydrologic contaminant transport model, *IGWMC Key No. 0693*, International Ground Water Modeling Center Database, Holcomb Res. Inst., Indianapolis, Ind., 1975.

International Ground Water Modeling Center, Mass transport models which are documented and available, *Rep. GWM 82-02*, Holcomb Res. Inst., Indianapolis, Ind., 1983.

Josephson, J., Protecting public groundwater supplies, *Environ. Sci. Technol.*, *16*(9), 502A-505A, 1982.

Keely, J. F., Chemical time-series sampling: A necessary technique, in Proceedings: Second National Symposium on Aquifer Restoration and Groundwater Monitoring, pp. 133-147, National Water Well Assoc., Worthington, Ohio, 1982.

Kimmel, G. E., and O. C. Braids, Leachate plumes in a highly permeable aquifer, *Ground Water, 12*, 388-393, 1974.

Knowles, T. R., GWSIM-II, *IGWMC Key No. 0680*, International Ground Water Modeling Center Database, Holcomb Res. Inst., Indianapolis, Ind., 1981.

Konikow, L. F., and J. D. Bredehoeft, Computer model of two-dimensional solute transport and dispersion in ground water, in *Tech. Water Resour. Invest.*, Book 7, Chap. C2, 90 pp., U.S. Geol. Surv., Reston, Virg., 1978.

Korver, J. A., The OGRE code: A two-dimensional numerical model of the transient flow of one or two compressible fluids through confined porous media, *Rep. UCRL-50820*, Lawrence Livermore Lab., Livermore, Calif., 1970.

Lallemand-Barres, A., and Peaudecerf, P., Recherche des relations entre la valeur de la dispersivite macroscopique d'un milieu aquifere, ses autres caracteristiques et les conditions de mesure, Etude bibliographique, *Bull. Bur. Rech. Geol. Min. Fr., 2nd Series, Sect. III, 4*, 277-284, 1978.

Ledoux, E., NEWSAM, *IGWMC Key No. 1450,* International Ground Water Modeling Center Database, Holcomb Res. Inst., Indianapolis, Ind., 1976.

Lippmann, M. J., C. F. Tsang, and P. A. Witherspoon, Analysis of the response of geothermal reservoirs under injection and production procedures, *Rep. SPE 6537*, Soc. of Pet. Eng., Dallas, Tex., 1977.

Marlon-Lambert, J., Computer programs for ground water flow and solute transport analysis, *Rep. N25090*, Golder Associates, Vancouver, B.C., 1978.

Moench, A. F., and A. Ogata, A numerical inversion of the Laplace transform solution to radial dispersion in a porous medium, *Water Resour. Res.*, *17*(1), 250-252, 1981.

Molz, F. J., J. G. Melville, A. D. Parr, D. A. King, and M. T. Hopf, Aquifer thermal energy storage: A well doublet experiment at increased temperatures, *Water Resour. Res.*, *19*(1), 149-160, 1983.

Montague, P., Hazardous waste landfills: Some lessons from New Jersey, *Civ. Eng.*, *52*(9), 53-56, 1982.

Muskat, M., *The Flow of Homogeneous Fluids Through Porous Media,* 763 pp., McGraw-Hill, New York, 1937.

Narasimhan, T. N., and P. A. Witherspoon, An integrated finite difference method for analyzing fluid flow in porous media, *Water Resour. Res.*, *12*(1), 57-64, 1976.

Nelson, R. W., Evaluating the environmental consequences of groundwater contamination, 2, Obtaining location/arrival time and location/outflow quantity distributions for steady flow systems, *Water Resour. Res.*, *14*(3), 416-428, 1978.

Nelson, R. W., and J. A. Schur, PATHS- ground water hydrologic model (Assessment of effectiveness of geologic isolation systems), *Rep. PNL-3162*, Pacific Northwest Lab., Richland, Wash., 1980.

Noorishad, J., and M. Mehran, An upstream finite element method for solution of transient transport equations in fractured porous media, *Water Resour. Res.*, *18*(3), 588-596, 1982.

Numerical Algorithms Group (NAG), *NAG Fortran Library Manual*, Downer's Grove, Ill., 1981.

Ogata, A., Dispersion in porous media, Ph.D. thesis, 121 pp., Northwestern Univ., Evanston, Ill., 1958.

Ogata, A., Theory of dispersion in a granular medium, *U.S. Geol. Surv. Prof. Pap. 411-I*, 34 pp., 1970.

Patterson, R. J., and T. Spoel, Laboratory measurements of the Strontium distribution coefficient K_d^{Sr} for sediments from a shallow sand aquifer, *Water Resour. Res.*, *17*(3), 513-520, 1981.

Perkins, T. K., and O. C. Johnston, A review of diffusion and dispersion in porous media, *Soc. Pet. Eng. J.*, *3*, 70-84, 1963.

Pickens, J. F., and G. E. Grisak, Finite element analysis of liquid flow, heat transport and solute transport in a ground-water flow system, 1, Governing equation and model formulation, report, Natl. Hydrol. Res. Inst., Inland Waters Direct. Environ. Canada, Ottawa, Ont., 1979.

Pickens, J. F., R. E. Jackson, K. J. Inch, and W. F. Merrit, Measurement of distribution coefficients using a radial injection dual-tracer test, *Water Resour. Res.*, *17*(3), 529-544, 1981.

Pinder, G. F., A Galerkin-finite-element simulation of groundwater contamination on Long Island, New York, *Water Resour. Res.*, *9*(6), 1657-1669, 1973.

Pinder, G. F., and W. G. Gray, *Finite Element Simulation in Surface and Subsurface Hydrology*, 295 pp., Academic, New York, 1977.

Prickett, T. A., T. G. Naymik, and C. G. Lonnquist, A random-walk solute transport model for selected groundwater quality evaluations, *Bull. 65*, 103 pp., Ill. State Water Surv., Champaign, 1981.

Raimondi, P., G. H. G. Gardner, and C. B. Petrick, Effect of pore structure molecular diffusion on the mixing of miscible liquids flowing in porous media, paper presented at Conference on Fundamental Concepts of Miscible Fluid Dispersion, Part 1, American Inst. Chem. Eng.-Soc. Pet. Eng., San Francisco, Dec., 1959.

Reisenauer, A. E., K. T. Key, T. N. Narasimhan, and R. W. Nelson, TRUST: A computer program for variably saturated flow in multidimensional, deformable media, *Rep. PNL-3975 RU*, Pacific Northwest Lab., Richland, Wash., 1982. (Also available as *U.S. Nucl. Regul. Comm. Rep. NUREG/CR-2360*, 1982.)

Robertson, J. B., Digital modeling of radioactive and chemical waste transport in the Snake River Plain aquifer at the National Reactor Testing Station, Idaho, *AEC No. IDO-22054*, open file report, 41 pp., U.S. Geol. Surv., Menlo Park, Calif., 1974.

Robertson, J. M., C. R. Toussaint, and M. A. Jorque, Organic compounds entering groundwater from a landfill, *U.S. Environ. Prot. Agen. Off. Res. Dev. Rep. EPA 660/2-74-077*, Washington, D. C., 1974.

Robinson, R. A., and R. H. Stokes, *Electrolyte Solutions*, 2nd ed., Butterworths, London, 1965.

Ross, B., C. M. Koplik, M. S. Giuffre, S. P. Hodgin, and J. J. Duffy, NUTRAN: A computer model of long-term hazards from waste repositories, *Rep. UCRL-15150*, Analytic Sci. Corp., Reading, Mass., 1979.

Runchal, A., J. Treger, and G. Segal, Program EP21 (GWTHERM): Two-dimensional fluid flow, heat, and mass transport in porous media, *Tech. Note TN-LA-34*, Adv. Technol. Group, Dames and Moore, Los Angeles, 1979.

Scheidegger, A. E., General theory of dispersion in porous media, *J. Geophys. Res.*, *66*(10), 3273-3278, 1961.

Schwartz, F. W., and A. Crowe, A deterministic-probabilistic model for contaminant transport, *U.S. Nucl. Regul. Comm. Rep. NUREG/CR-1609*, 1980.

Science Applications, Inc., Tabulation of waste isolation computer models, *Rep. ONWI-78*, Off. of Nucl. Waste Isolation, Battelle Memorial Inst., Columbus, Ohio, 1981.

Segol, G. A., A three-dimensional Galerkin finite element model for the analysis of contaminant transport in variably saturated porous media: User's Guide, Dep. of Earth Sci., Univ. of Waterloo, Waterloo, Ont., 1976.

Stehfest, H., Numerical inversion of Laplace transforms, *Commun. ACM, 13*(1), 47-49, 1970.

Terzaghi, C., Principles of soil mechanics: Settlement and consolidation of clays, *Eng. News. Rec.*, *95*, 874-878, 1925.

Thomas, S. D., B. Ross, and J. W. Mercer, A summary of repository siting models, *U.S. Nucl. Regul. Comm. Rep. NUREG/CR-2782*, 1982.

Tracy, J. V., User's guide and documentation for adsorption and decay modifications to the U.S.G.S. solute transport model, *U.S. Nucl. Regul. Comm. Rep. NUREG/CR-2502*, 1982.

U.S. Environmental Protection Agency, Handbook for monitoring industrial waste water, Washington, D. C., 1973.

U.S. Environmental Protection Agency, Summary report: Gas and leachate from land disposal of municipal solid waste, 62 pp., Solid and Hazardous Waste Res. Lab., Natl. Environ. Res. Center, Cincinnati, Ohio, 1974.

U.S. Environmental Protection Agency, *Waste Disposal Practices and Their Effects on Ground Water: The Report to Congress*, Washington, D. C., 1977.

Van Genuchten, M. Th., One-dimensional analytical transport modeling, in Proceedings: Symposium on Unsaturated Flow and Transport Modeling, *Rep. PNL-SA-10325*, Pacific Northwest Lab., Richland, Wash., 1982. (Also available as *U.S. Nucl. Regul. Comm. Rep. NUREG/CP-0030*, 1982.)

Van Genuchten, M. Th., and W. J. Alves, Analytical solutions of the one-dimensional convective-dispersive solute transport equation, *U.S. Dep. of Agric. Tech. Bull. 1661*, 149 pp., 1982.

Wang, J. S. Y., C. F. Tsang, and R. A. Sterbentz, The state of the art of numerical modeling of thermohydrological flow in fractured rock masses, *Environ. Geol.*, *4*, 133-199, 1983.

Warner, J. W., Finite element two-dimensional transport model of ground water restoration for in-situ solution mining of uranium, Ph.D. thesis, Colorado State Univ., Fort Collins, 1981.

Yeh, G. T., and D. S. Ward, FEMWASTE: A finite-element model of waste transport through saturated-unsaturated porous media, *Rep. ORNL-5601*, Oak Ridge Natl. Lab., Oak Ridge, Tenn., 1981.

Zienkiewicz, O. C., *The Finite Element Method*, 3rd ed., 787 pp., McGraw-Hill, New York, 1977.